# Hands on the Wheel

## A Human-Centered Guide to AI in the Classroom

Douglas Kiang

AI Forward Press

## Dedication

*For my mother and all the other English teachers, past and present, who fostered in me a love of the written word.*

# Contents

# Foreword

I first met Douglas Kiang in 2018, when we both served on the College Board's development committee for AP Computer Science Principles. I quickly grew to admire Douglas as an educator and as a colleague who thinks hard about things that are hard to think about, never sparing his own ideas the same scrutiny he gives everyone else's. His point of view is always deeply considered, egoless, and pragmatic; his suggestions are concrete, actionable, and grounded in real-world experience. What I value even more is his constant wonder at what new technology can do in a classroom.

I know this is beginning to read like a letter of recommendation. I write it because it describes the voice of Douglas Kiang and this is exactly the voice I read on every page of this book.

I have taught computer science at Columbia University for over twenty-five years. When I began, there were still people panicking about what the World Wide Web would do to education. The Web changed how students did research; it didn't eliminate the need to do it. Assignments changed, syllabi adjusted, teachers adapted. Then came smartphones, free massive open online courses, cheap and ubiquitous homework outsourcing, and finally a global pandemic. Each one disrupted our approach to teaching in its own way, and each time, a dedicated educator was compelled to reexamine their approach and adapt.

Funny thing is, the result usually improved how we teach. Web search let me build assignments that demanded more research, pushing students to find the most relevant sources in a sea of related articles. Smartphones let me run live polling and roll call in large lectures. Online outsourcing pushed me toward a cloud-based IDE that can replay every keystroke a student makes when building a project, which gives me data that improves every instance of my course. The pandemic taught me that using a TA to monitor an online discussion board during lecture draws out questions from students who'd never raise their hand in person. I use screencasts far more effectively now, after two years of living in two dimensions. Everyone in education has a version of these stories.

Is it different this time? Yes. I think AI will be more disruptive than everything I just listed, combined. Will we adapt? Of course we will. As Douglas points out more than once in the pages that follow, AI is a tool that ultimately requires us to reexamine what we teach, how we teach it, and why we teach it. We cannot ignore it. Hands on the Wheel isn't about wrestling control back from the machine, or handing it over entirely. It's about using AI to move forward while still maintaining control over where we're going and how we're getting there. Like Douglas, I'm not an AI evangelist, I'm a pragmatist. This may not be the future we expected, but here we are, and we have work to do.

This book won't try to convince you that AI is the best thing that's happened to education, or the worst. It offers something more useful: an honest account of what actually happens when these tools are thoughtfully used in an assignment, a lab, a classroom discussion, or even a faculty meeting. Douglas writes with the warmth and humility of someone who has done the work, hit the setbacks, and turned enough of them into breakthroughs to have real advice to give. His clear understanding of the technical fundamentals and long history of innovation in computer science education imbue him with a tremendously valuable point of view. I have had many conversations with Douglas about AI over the past few years, and I still bring what I learn from them into my own classes. This book is full of that same kind of insight: practical enough to use tomorrow, honest enough to trust. Try what resonates with you; find out what works in your classroom.

I don't think it's an overstatement to say that how this moment in secondary education is handled will shape how an entire generation understands their own capability. Students arriving at college today have already been using AI for years! Think about that. Handled badly, AI becomes a way for students to avoid developing the kind of thinking that they will desperately need in college. Handled well, as Douglas shows throughout this book, it becomes a way to amplify student learning, empowering students to attempt more than they ever did before. This book is a genuinely useful, clear-eyed guide to doing the second thing on purpose.

I'm grateful Douglas asked me to write these opening words, and thrilled with what he's accomplished in *Hands on the Wheel*. Whether you are a new teacher trying to write your first AI-era syllabus or a twenty-year veteran wondering what just happened to your classroom, you are in good hands.

*Adam Cannon*
Teaching Professor of Computer Science
Columbia University

## How This Book Was Built

Somewhere around the third chapter, the AI told me my best argument was weak.

I had written a paragraph I was proud of, about why teachers should create with AI before they teach with it, and I asked the model to push back the way a skeptical department chair would. It found the soft spot in about four seconds. The claim leaned on an example that didn't actually prove what I said it proved. I sat there annoyed, because it was right, and then I rewrote the paragraph until the example actually backed up what I'd said.

That exchange is this book in miniature, which is why it belongs up front instead of buried in a footnote.

Everything in these pages started with me. The classroom scenes happened in my classroom. The students are real, and the mistakes are mine: the ones I made then and the ones I'm still making now. Many of the chapters began as essays I wrote by hand over the past year for my blog hosted at kiang.net, drafted late at night with no machine in the room.

And I built the book itself using the method I lay out in Part II, "Use It."

I gave the model my own writing as context first, a principle I call Always Build Context, and I directed everything from there. I set the vision for each chapter and decided what the book was for. I used the tool to sharpen prose that had gone slack and to run structural diagnostics on chapters I was too close to, to read clearly. I made it interrogate the arguments I wanted to believe but hadn't tested. When it handed me something good, I rewrote it in my own voice, and when it handed me something that sounded like everyone and no one, I threw it out.

I want to be clear about that, because the temptation with AI is to let it do more and call the result yours. It doesn't work that way. The more the model gave me, the more of my own judgment I had to put back in to make the writing mine. That is the maxim underneath this whole book: The more AI gives you, the more of yourself you need to put in.

Finally, I hired a copy editor to provide human feedback, and an artist to design the cover. What they contributed to this finished work was unique and essential and I would not do a project like this without involving humans in the process, no matter how advanced the tools.

If you run a school, you are being asked to make decisions about a technology whose inner workings can feel like a black box, and that is a hard place to lead from. This book, the one in your hands, was made with the same tools I would teach your faculty to use, by a teacher who stayed in charge of every word. I'm not describing AI from across the room. I ran the experiment on myself first, chapter after chapter, and wrote down what actually happened, wins and failures both.

Everything I claim here, I checked myself, and every story is true. The voice is mine, and so were the hands on the wheel.

# Part I: FACE IT

*Why no school opts out*

# 1

## We Didn't Ask for This

We have all seen the movie, or one like it. Huge alien ships are parked over every city on Earth, our governments paralyzed, debating what to do, shocked into inaction. Among the human population, there are those who form the resistance. Some of us can identify with those heroes: the ones who fly the fighter jets, who make courageous speeches, who rally the rest of humanity to fight back against this alien, existential threat.

Artificial Intelligence (ChatGPT, DeepSeek, Gemini, and their kind) landed on our doorsteps like an alien invasion. Many are fighting back. I will be honest with you. I am not in the Resistance.

I am what you would call a collaborator. I am the guy holding a "Welcome to Earth" sign at the landing zone. I brought bottled water and the Wi-Fi password. I say please and thank you. I am the one saying, "Let me take you to our leaders. We can use my car."

I know this is not heroic. I think of myself as pragmatic. Maybe I'm trying to make sure AI likes you and me just a little more than everyone else. Because if this ever turns ugly, I want at least one of them pointing at me saying, "Not that one. He lent us his car."

The reason I'm writing this book is to invite you, if you're willing, to explore this new technology along with me. I'm a teacher in Silicon Valley, the epicenter where so many of these AI tools are being developed.

Than Healy, our head of school, pulled me into a conversation about a year before I started writing this book. At the time, most of the talk in education was about restricting AI. Detecting it, keeping it out of student work. Schools were writing policies as fast as they could, and most of those policies boiled down to the same word: *Don't.*

Than wanted to flip that entirely. He asked a simple question: "What if we had a course called You Must Use AI in This Course?"

I asked Rachel Blumenthal, our English department chair, if she was up for an adventure, and to my delight she said yes immediately. We spent a year planning what became *Literature in the Age of AI*, a semester-long English elective for juniors and seniors (ages sixteen to eighteen) at Menlo School, the independent school in Silicon Valley where I teach. We called the approach AI Forward, and the idea was simple: Guide and shape students' use of AI through positive and productive work rather than through a list of constraints. The rest of this book is about what we learned. But this chapter isn't about the course itself. It's about why we had no choice but to build it.

## The speed of the thing

I say the same thing about artificial intelligence (AI) to teachers every time I give a talk, because it's true and because it's important to hear someone say it out loud. This wasn't a carefully rolled-out initiative. There was no pilot program and no planning time. This makes the gold rush look like a piñata grab.

AI was not an educational initiative. It was a product launched by technology companies who view education, if they think about it at all, as an afterthought. And then teachers were expected to figure it out.

There's a name for what makes AI's introduction different. Justin Reich, who directs the MIT Teaching Systems Lab, and the researcher Jesse Dukes call generative AI an *arrival technology*, to set it apart from the *adoption technologies* schools are built around. An adoption technology gets evaluated and piloted before it ever reaches a classroom, if it reaches one at all. An arrival technology skips every step of that. It shows up in students' hands whether or not anyone in charge decided to let it in, which means the question schools are used to asking—*Should we adopt this?*—is one AI has already answered for us.[1]

I show a video in my keynotes that tends to ruin people's afternoons.[2] The education researcher David Wiley posted it, and it's only about three-and-a-half minutes long. In the video, a student logs in to Canvas, the learning management system many schools use for coursework, and has an AI agent find the next assignment, read the prompt, draft a full response in about twenty seconds, and submit it. The student never reads the assignment. The student never reads the response. The AI does the whole thing, start to finish, and at the end it asks politely, "Would you like me to submit?" Done.

The assignment, by the way, was about how AI impacts education. I wish I were making that up.

Wiley's comment afterward was pointed: "And for those who are concerned that this video shows students how to use generative AI to do their homework for them... They already know how to do it. It's only faculty who don't know how."

Today, AI is the worst AI you will ever use. The version you're looking at right now, the one that already feels uncomfortably capable, is the slowest and least powerful version you will ever

---

1. Justin Reich and Jesse Dukes, "The future of education technology after the arrival of ChatGPT," *Phi Delta Kappan* 107, no. 3–4 (2025). Reich directs the MIT Teaching Systems Lab, and the distinction between technologies schools adopt and technologies that simply arrive runs through his broader work on why so much classroom technology disappoints.

2. If you want to feel the same unease I felt watching the Canvas video for the first time, search for "using ChatGPT's agent mode to autonomously complete online homework" and brace yourself. The three-and-a-half minutes will feel like ten.

encounter for the rest of your career. Second-semester AI will be different from first-semester AI. Next year's AI will make this year's look like a flip phone. The pace is not slowing down, and no school is going to wait it out.

## The grief is real

So if you're frustrated, you should be.

If you're grieving assignments you loved, units you spent years refining, activities you can't give anymore in their current form, that grief is real. I've watched colleagues move through something very close to the stages of grief. Denial: I'm just going to pretend this doesn't exist and keep doing things the same way. Anger. Blame. And then, slowly, something that looks like acceptance, though most teachers I know would call it something more honest, like resignation with a plan.

Everyone is moving through this adjustment at different speeds, and that's okay. What is not okay is deciding not to move through it at all.

I keep telling administrators and department heads the same thing. You can decide not to use AI. You can decide the whole thing is overhyped and that the fundamentals of good teaching haven't changed (and you'd be partly right about that). But you cannot decide that your students aren't using it. They are. They were using it before you heard of ChatGPT. They are using it even when they don't intend to use it. It's baked into Google searches. It's embedded in their workspace. It's in their browser as a Grammarly plugin. When they turn in a paper, students may honestly not remember exactly how much of their writing is theirs. One of my tenth graders put it to me with the kind of clarity only a sixteen-year-old can manage: "If it's good, I say I wrote it. If it's bad, I say it was AI."

## We've done this before

We didn't ask for this. That matters.

AI arrived in schools the way the COVID pandemic arrived: without warning and without a playbook. I was interviewing for my job at Menlo the very week the school decided to close, one of the first schools in the country to shut down, and within days every teacher in the building was teaching through a screen. Nobody had trained for it. Nobody had asked for it.

The teachers who thrived during the pandemic weren't the ones with the best tech skills. They were the ones who leaned on what they already knew about reaching a kid under bad conditions. They figured out that a thirty-minute video chat, even an awkward one, could be enough to make a student feel known. That their presence, even through a webcam, was still valuable. They adapted not because they had a playbook but because they understood something about kids that no technology can replicate.

AI presents the same challenge. It is unchosen and disruptive, and it is potentially useful, but only if we bring our expertise to it rather than waiting for the tech companies to design classroom solutions for us. They're not going to. They're building tools for productivity and profit. Education is something we have to figure out ourselves, the way we always have.

## The gate just fell

The COVID-19 pandemic caught us all by surprise. It began with an abrupt shift in the ways we teach, work, and communicate. It involved a rapid, tumultuous period of adjustment. But it ended with our adapting, and surviving—even, in some cases, thriving. Adapting quickly is something we teachers know how to do. We have done it before. There are reasons to be excited about the arrival of AI that have nothing to do with coping. That's why I'm writing an AI-forward book instead of an AI survival guide.

Mia took my intro computer science class. Two semesters. She learned enough to write small programs and to read an error message without panicking. That's the whole foundation. It's the foundation of skills thousands of kids get and then too often set aside, because the distance between "I passed intro CS" and "I built an app people actually use" used to be years of additional coursework that most students never take.

She didn't wait to take years of CS. She sat down with AI and built an app. Not a toy and not a class assignment, but a real thing that is on the App Store right now. It helps people with glaucoma take their eye drops.

Her grandfather had glaucoma and lost his sight. There were several reasons, but one of the biggest was that he didn't understand how much the daily drops mattered until it was too late for them to matter. Mia grew up watching that happen.

Glaucoma is quiet in the early years. Patients don't feel any better for taking their medication, the drops themselves are unpleasant, and there is no cure waiting at the end of the routine, so a lot of people simply stop. Roughly half are no longer taking their medication a year after diagnosis. There is no warning when they quit. The sight goes slowly, and by the time it's obvious, it's gone.

So she built the vital daily routine into an app. The app reminds you when a dose is due and records whether it happened. Stacked around that is everything else the routine actually asks of a person: step-by-step instructions for getting a drop into your own eye, a symptom log, refill reminders, short lessons about the disease, a calendar of every dose you've taken, and a PDF you can export and hand to your ophthalmologist at your next appointment.

Think about who built that and why. A teenager, handed a tool that could write her almost anything, used it on the thing that took her grandfather's vision.

The AI wrote much of the code, under Mia's direction. For some CS teachers, this might nag at them because she didn't write all of it.

But here is what she actually did. She had the idea, which no tool gave her. Two semesters of CS were enough to keep her hands on the wheel: enough to keep the app headed where she wanted it and to catch the AI when it broke something. And she decided what the thing was for, which is the part that matters most and the part a model will never decide for you.

Then she did the part that has nothing to do with code at all. She cold-emailed glaucoma specialists and researchers until a few of them wrote back. She met with clinicians at Duke, UC San Diego, and Stanford, listened to what they told her was wrong, and rebuilt the app more times than she can count. A physician at the Stanford Byers Eye Institute is now mentoring her and preparing to put it in front of real patients. No model did any of that for her.

I'm not going to tell you AI turns every intro student into a software engineer, or that two semesters of anything is enough on its own. You've read enough of this book already to know I don't believe that. What I am claiming is narrower and it holds up better. The part that used to keep kids out, the years of technical training you needed before you could build something real, is no longer the only door in. A student with a foundation and a reason can now reach a mile past where that foundation alone would leave her. What separates the work worth using from the work worth forgetting isn't whether she could write every line herself. It's what she brought to it. Mia brought a grandfather who went blind and an exact sense of how ordinary it looks when a person quietly gives up on a treatment. No tool was ever going to hand her that.

This is the most hopeful thing I know about AI in the classroom. For most of my career, the kids who got to make ambitious things were the ones who came in with a head start in STEM: the four years of CS electives, or the parent who could pay for coding lessons. The coding gate sorted those kids in and sorted a lot of other kids out. When the distance from foundation to finished thing collapses, the sorting changes. The kid who only ever took the intro class can build the thing she's been picturing for a year.

The kid who was told in sixth grade that he wasn't a "STEM kid" can find out whether that was ever true.

The fact that we have lowered the barriers to participation does not mean what these kids make is any less real, any more than a first novel is less real because a tool helped carry it. More kids clearing the barrier just means more kids clearing the barrier. The catch, and there's always a catch, is that a room full of students who can all build things is also a room where the average thing is suddenly everywhere. What stays rare is the student who brings something the average can't. Mia did. Most people use AI to shorten their to-do lists. She used it to build an app that saves the eyesight of other people's grandparents, mothers, and fathers. That instinct, the one that decides what a technology is for, is exactly what the rest of this book is trying to protect in you, so that you can protect it in your students.

## Your instincts are better than you think

The good news (and I mean this, not as a pep talk but as a factual observation from working with thousands of teachers over the past fifteen years) is that your professional instincts are already well suited to the challenge of teaching with AI.

Teachers are the original prompt engineers. We've been writing clear directions, setting context, defining success criteria, and iterating on assignments for decades. Prompts are mini-assignments. When you write instructions for a class of ninth graders and anticipate every possible misreading, you are practicing exactly the skill that makes someone effective with AI. When you look at student work and can tell something is off before you can articulate why, you are doing the kind of qualitative evaluation that no AI detector on the market can match.

We know our students. We know what real thinking looks like, and we can tell when it's missing, even when the software can't.

Your instincts transfer. That part is real. But instinct alone isn't enough, and this is where I reveal something this book is going to keep coming back to.

## You can't teach what you won't touch

The one rule I've landed on after two years of teaching with AI, building with AI, writing a trashy romance novel with AI, running a faculty professional development series, coaching other schools, and making more mistakes than I'll probably admit even in this book, is this: You cannot teach something you are unwilling to do yourself.

I don't mean you need to become an AI expert or understand how neural networks function. I mean you need to sit down with one of these tools, give it twenty minutes and a real task from your own work, and pay attention to what happens. Not a toy task. Not "Write me a haiku about fractions." Something you were already going to do this week, like a rubric, a parent email, or feedback on a paper.

Because until you do that, your opinions about AI in education are theoretical. And your students, who use these tools as casually as they use a search engine, will know the difference.

I started this way myself. For most of 2023, I used ChatGPT maybe once a week. I'd lived most of my career managing just fine without it, and the whole thing seemed like a cool trick, impressive but not essential. Then I started asking myself a different question every time I sat down to work: "Is this something AI can do?" Not "Should AI do this?" or "Is it ethical for AI to do this?" Both of these are important questions I'll get to later. Just: *Can it?*

The answer was *yes* far more often than I expected. And the interesting part wasn't the tasks AI handled well. It was the ones where it fell short in ways I could see but couldn't have predicted without trying. Those gaps, the places where the tool is fluent but wrong, or fast but missing something a human would catch in a second, are where the teaching lives. You can't find them by reading about AI. You find them by using it.

That's what this book is going to ask you to do. Not to become an evangelist, and not to upend your curriculum overnight. I'm going to ask you to *create* with AI before we talk about how your students should, because the judgment you need to guide them is judgment you can only develop through your own hands.

> *You can't teach what you won't touch. Your opinion about AI in your classroom should come from your own hours in the tool, not from a headline.*

## What's coming

The rest of Part I is about what happens when you look honestly at the assignments AI has already broken or destroyed, and treat the damage as a design problem rather than a discipline problem. Part II asks you to make something, in your own subject, so you understand what the tool can and can't do before you bring it to students. Part III is about classroom practice, including the part I care most about, which is what it looks like when we lower many of the barriers to creativity. Part IV is the critical-thinking core: the honest, hard questions that many books about AI in education skip. And Part V is about moving a whole school to adopt an AI-forward approach.

Every so often you'll run into a short box labeled "Under the Hood." Each one explains the machinery behind something the tool just did, like why it drifted toward the most generic answer, or why it sounded completely sure while being completely wrong. You don't need to read these technical sections to teach well. If the mechanics hold no interest for you, skip reading them without guilt.

I should tell you now that this book is not neutral. I believe AI is going to change what it means to teach as profoundly as the internet did, and possibly faster. I also believe that teachers are the people best equipped to make that change go well, because we are the people who know kids and who have been adapting to unforeseen circumstances for as long as schools have existed.

But I also believe the harm AI creates is real and the story of this transformation is far from settled. Some of what these tools cost to build is indefensible, and nothing in the rest of this book makes that part go away.

We didn't ask for this. But here's what I keep coming back to. The same tools that landed on us uninvited are the ones about to let a lot more kids make the things they were always capable of making. We're the people standing between those kids and the flat, average, non-Mia version of that promise. That isn't damage control. That's the best job in the building.

### Run It: Give it your next real task

**Yourself.** Take the next thing you were going to do for work and give it to AI instead. Don't choose a hypothetical task. Choose the actual next thing: a rubric revision, an email to a parent, a set of discussion questions, feedback on a draft. Give it twenty minutes and your honest attention, then write down three answers. What did the AI do well enough that you'd use it again? Where did you have to fix something, and what did you know that it didn't? What surprised you? If the answers change how you think about AI in your classroom, good. If they don't, you're out twenty minutes, and you'll have an opinion grounded in something real instead of anxiety.

**Students.** Ask your students what they'd build if the technical part were already handled. Not as an assignment, just a question, the way Mia had a year of picturing her app before she ever wrote a line of it. Give them ten minutes to describe the thing they'd make and, more importantly, why it has to exist. You are not teaching them to build anything yet. You're finding out what they'd point a tool at, which is the one thing the tool can't decide for them, and the thing you'll spend the rest of the year protecting.

**Colleagues.** Bring the twenty-minute exercise to your next department or faculty meeting and have everyone do it at once, each with a task from their own subject. Then go around the room. The disagreement about what the AI handled well and where it fell down is the most useful professional development conversation most departments haven't had. The rule that holds the whole thing together: Nobody teaches with the tool until everyone in the room has touched it themselves.

# 2

## It's a Design Problem, not a Cheating Problem

A colleague of mine spent years perfecting a game design assignment. Students would pick a board game, play it, and then write an analysis of the game dynamics: **push your luck**, **risk versus reward**, or **cooperative versus competitive tension**. It was a beloved unit. Kids loved it. She loved teaching it.

Last fall, she gave the same assignment. A student turned in a flawless analysis in twenty minutes: too polished and too fast. My colleague ran it through every AI detector she could find. Inconclusive. She couldn't prove anything. And the worst part? The student probably didn't even think he was doing anything wrong. He'd used the tool on his phone the same way he'd use a calculator.

That unit, the one my colleague had spent years building, was done. Not because it was a bad assignment, but because the world had changed.

The instinct in a moment like that is to go after the student. Or to go after the tool, ban it and punish anyone caught using it. That instinct is understandable, but I think it's aimed at the wrong target.

The assignment broke. The student just happened to be there when it did.

## The assignment is the variable

If an AI can ace your assignment cold, with no context, no prior reading, and no lived experience in your classroom, then the assignment was testing something the AI can do trivially. That's a design flaw in the assignment, not a moral failing on the part of a sixteen-year-old.

I tell this to teachers gently, because I know how it sounds. Nobody wants to hear that their best assignment, the one they'd been refining for a decade, is now the equivalent of asking students to copy an entry out of the encyclopedia. But that's what happened to a lot of our work in November 2022, when ChatGPT burst onto the global stage, and pretending otherwise doesn't help us or our students.

The good news is that fixing the assignment is almost always easier than policing the behavior, and it produces better learning. My colleague redesigned her game design unit. Now students play four or five board games, taking notes. Then they have AI write the individual analyses. And then they write their own meta-analysis comparing all of them: what they agreed with, where the AI-written analyses were wrong, which game dynamics they personally enjoy and why, what makes one game pull them in and another leave them cold. The assignment asks for the one thing AI can't fake, which is a particular kid's actual experience and honest reflection.

It's a better assignment than the original. She'll tell you that herself.

## You can't cut this tree down

We can't cut this tree down. Believe me, people have tried.[1]

---

1. This is the short version of a story I tell in full at the end of the book, about a tree on a Hawaiian volcano that everyone thought was growing terrible oranges.

Schools that try anyway learn the same lesson eventually. A student of mine told me about a friend at a school I'll call Bann Academy. Bann bans AI on all schoolwork and promises to expel anyone caught using it. To date, nobody has been caught, because the students have gotten so good at coaching AI to sound like them that the detectors can't tell the difference. One student at Bann told my student she spends more time fooling the AI detector than she spent writing her first draft.

When I first heard that, I was appalled. Then I thought, wait. Is Bann onto something? Those students are actually developing a useful future skill: the ability to recognize what AI-generated writing sounds like and to push it toward something personal. We want our students to learn the same thing. Both schools are fumbling toward the same future, from opposite directions.

The same skill points in two directions. Aimed at developing your own voice, coaching AI to sound less generic and more like you, it's worth teaching, and we teach it in the open. Aimed at slipping past a detector, it's the identical skill spent entirely on concealment, and the student who masters it has learned to hide rather than to write. Bann's students are getting good at the right thing for the wrong reason, and the reason is what their school built for them.

## Club Scout and the wrong problem

I like to challenge my students to create apps that solve real problems. I am always hopeful that they will use their skills to make a difference in the world, and change it for the better. I used to ask them to be problem detectives: Find a problem that the world is facing, and invent a solution.

One day, two of my students came up to me, beyond excited. Mr. Kiang, they said, we've found a perfect problem to solve. I said great, what's the big problem? They said: When we go to the golf course with my mom, she never knows the right club to use.

So we're going to build an app that measures the wind and the distance and tells her, use a nine iron.

I said, this was the biggest problem you guys found? They said, Mr. Kiang, this is a big problem. Trust us. So they built it. They called it Club Scout. They built it, it worked, and it was genuinely well crafted. They learned to code. They satisfied every requirement on my rubric. The problem was that "helping my mom pick a nine iron" didn't quite crack the list of the world's most pressing problems.

The next year, I started using the UN Global Goals, the seventeen biggest challenges facing the world, as the frame for student projects. These are big challenges like health and well-being, education, and climate action (choosing the right golf club must have been number eighteen because it didn't make the cut).

Club Scout was technically impressive and completely beside the point. AI will do the same thing for you if you give it an assignment without purpose behind it. It will deliver polished, competent solutions to problems that don't actually matter, and it will assure you that your idea is brilliant in every case. The ability to look at excellent execution and ask "But is this worth doing?" is a teacher's instinct, and no prompting framework teaches it.

## My call, my craft, my charge

Some years ago, before any of this, I was carrying a full teaching load, advising a school club, consulting for Apple, running workshops for EdTechTeacher, and sitting on the development committee for the AP Computer Science Principles exam. On top of all of it, three nights a week, I taught driver's education. Something had to give, and driver's ed was the obvious thing to cut. It paid the least of anything I did, and it had nothing to do with the rest of my career. So one evening I sat down to quit, and I found that I couldn't.

It took me a while to work out why. Around that time I read Daniel Pink's *Drive*, which argues that three things make work feel worth doing: autonomy, mastery, and purpose.[2] I realized that of everything on my plate, driver's ed was the one place I had all three at once.

The autonomy was almost total. The state gave me thirty hours of required content and left the rest to me. Nobody reviewed my lesson plans or told me how to run the room. We built model engines so students could see what was happening under the hood. We watched old safety films and picked apart what they got wrong. The final exam was my own.

I had also been driving for more than thirty years by then, and there was very little a person could do on the road that I hadn't seen or done, and couldn't teach my way through. That was the mastery, and it had taken decades to earn.

The purpose was the part that really mattered to me the most. I cared deeply about making sure no one would ever get the dreaded call in the small hours of the morning saying that one of our high schoolers had died behind the wheel.

Over the years since, teaching my own subject, I have come to name those three things in my own words. **My call. My craft. My charge.** My call is the knowledge that the work is mine to shape. My craft is the skill to do it. My charge is the person on the other end who is counting on me to get it right. When a project truly gives a student all three, they guard it. They will not hand it over wholesale to a machine. And I arrived at this well before there was a machine to hand it to.

I use those three phrases now as a test of the assignment rather than the student. If a student can say all three about the work I've given them, I don't have a cheating problem. If they can't say even one, that's a problem with the assignment, not the kid.

A few years ago a parent came to my computer science room with a problem that had nothing to do with grades. I'll call him Paul. His wife was losing her voice to ALS, a neuromuscular dis-

---

2. Daniel H. Pink, *Drive: The Surprising Truth About What Motivates Us* (Riverhead Books, 2009).

ease that slowly takes away control of the body's muscles, and the first thing it took from her was speech. He wanted to know if my students could build an app to let her write from her bed on an iPad, and Paul would get an alert on his phone in the next room so he could read what she was trying to say.[3]

The kids said yes immediately.

Paul came to class in person and told them about his wife, about ALS, about what it's like when the person you love most in the world can no longer tell you what she needs. The students leaned in hard, the way teenagers do when they sense something real is at stake. As part of the design thinking process, every one of them took a twenty-four-hour vow of silence so they'd know what it felt like to lose your voice. If you know my kids, twenty-four hours of silence was, let's say, an extraordinary act. They carried the prototype around and used it to communicate with anyone who would normally hear them speak, and they came back with insights we never would have gotten from a whiteboard session: problems with the design that only surface when you actually have to use it to communicate.

I told them to use AI freely. We were racing against the clock, and these kids had already taken our intro course, so speed mattered more than pedagogy.

What they found surprised all of us. AI kept introducing errors, and when they asked it to fix one thing, it would often rewrite whole sections of working code and break what had been fine. They spent more time cleaning up after the AI than they would have spent writing the thing themselves. The interface needed to be usable by someone with very limited motor control, which meant it had to be simple, yet AI kept trying to make it more feature-rich when the entire design problem was restraint. Eventually they said it plainly: We care too much about this to trust it to AI. We're just going to do it ourselves.

---

3. The family in this story has been anonymized. The illness, the app, and the loss are real.

The project gave those students call, craft, and charge at once, the way driver's ed had given them to me. The call was theirs, every real decision about how the app should work. The craft was theirs, earned in the intro course they had already taken. And the charge was a real person, and time was of the essence. No assignment I could build from scratch or self-imposed deadline will ever manufacture that.

A few months in, Paul's wife passed away. He came back to our classroom, and it was a hard morning. The kids were crying. I was crying. Paul told us that if he was being honest, he'd known since August that it was probably too late for her. But the work could still matter, he said. It could help other families. He asked the students to keep going.

They did. Over winter break, with no grade attached and nobody telling them to, they set up a chat channel and kept sending code back and forth. The app is now headed toward distribution to families everywhere, not just people with ALS but anyone who has lost the ability to speak. It will give a lot of people their voice back.

People ask me all the time why we should still teach skills that AI can do. That class is the best answer I have. When the work matters enough, when a student is genuinely invested in something real, they need to be able to do it themselves, and they will refuse to just hand it over to a machine. My students didn't reject AI because they were against technology. They rejected it because they had the skills to do more and cared too much to do less. If a student can't say *my call*, *my craft*, *my charge* about the work I've handed them, the fault is in the assignment.

## What the redesign looks like

Here are five assignments that broke and how we rebuilt them. Seeing the transformation a few times makes it easier to spot in your own assignments.

**Game analysis → Meta-analysis.** The old assignment: Pick a board game and analyze its dynamics. AI does that in thirty seconds. The redesign: Play four or five games, have AI write the individual analyses, then write your own comparison of those analyses, focusing on what kinds of game dynamics you personally enjoy and why. The graded work now captures the sort of judgment AI can't do for you.

**Write-all-code → Explain-and-defend code.** The old assignment: Students write every line of code from scratch. AI can now do that in seconds, which is the problem. The redesign: Students may use AI-generated code, but they must test it, modify it, explain every line, and defend their choices. Assessment shifts from production to understanding.

**Independent first draft → Documented AI process.** The old assignment: Students independently produce a first draft. The problem is that AI can generate a complete draft immediately. The redesign: Students use AI as part of the writing process and document what prompts they used, what revisions they made, what they accepted and what they rejected, and why. You evaluate their judgment as the primary benchmark of quality.

**Problem solving → Error archaeology.** The old assignment: Solve a set of complex multi-step calculus or statistical probability problems, showing your work. The redesign: The instructor provides students with an AI-generated solution to a complex problem that contains a subtle, logical mathematical flaw (a hallucination or an invalid assumption). The students must find the error, explain why the AI made that specific logical leap, correct the math, and then write a guide for a human peer on how to avoid making that exact same mistake.

**Formal lab → Real-world troubleshooting.** The old assignment: Perform a standard titration or PCR (polymerase chain reaction) lab, record the numbers, and write a formal lab report discussing the expected chemical/biological mechanisms. The redesign: Students conduct the physical experiment and intentionally introduce an error or use an unexpected variable. They feed their raw, messy lab data into the AI and ask it to write the discussion section. The student then acts as a peer reviewer. They must annotate the AI-generated report, highlighting where the AI forced the data to fit standard textbook expectations, where it failed to account for real-world environmental anomalies, and how the physical experimental setup could be redesigned to eliminate the noise.

The pattern across all of them is the same. Let AI have the part it can do trivially, and put the grade on what still requires a human. This is what lowering the barrier looks like inside a single assignment. The execution got cheap, so the judgment is what we grade.

## Designing *for* the student, not *against* the tool

There's a version of "redesign your assignments" that sounds like blame: You did something wrong. Your assignment was bad. Fix it.

That's not what I mean. The assignments that broke were good assignments, many of them great, and they broke because the world changed underneath them, the same way a lesson plan built around looking things up in an encyclopedia would have broken when Google showed up. The design problem isn't that you failed. It's that the floor moved.

What I'm encouraging you to do is what good teachers have always done when conditions change: Look at the assignment, figure out what it was really testing, and rebuild it so the thing that matters is still the thing the student has to do. The meta-analysis is better than the analysis. The documented AI process tells you more about a student's thinking than a polished product ever did.

And when you can't redesign it, because sometimes you can't, run the test. Can the student say *my call*, *my craft*, *my charge* about the work? If yes, you probably don't have a problem. If no, you might. The place to look first is the assignment, not the syllabus.

## Run It: Give the assignment to AI first

**Yourself.** Pick one assignment you're planning to give this term. Before you hand it to students, hand it to AI. Paste the prompt exactly as written, with no special context, nothing a student wouldn't have on a Sunday night. If what comes back would earn a passing grade, you have a design problem, and now you can fix it before a student finds the same gap between assignment and effort. Look at what the AI got right and ask whether that was the part you were trying to assess.

**Students.** Do the same thing, but with your students watching. Take an assignment they've already done and generate an AI response to it live, on the screen, in front of the class. Then hand the grading to them. Ask them to mark it against your rubric and argue for the score. The conversation you want is the one where they can't agree: The places where half the room says it passes and half says it's hollow are exactly where the real skill lives, and now they can name it. End by asking them to redesign the assignment so the AI version wouldn't earn full marks. They are usually better at this than we are, and they will tell you things about your assignment you didn't know.

**Colleagues.** Bring one shared assignment to a department meeting and run the same test together, out loud. When you disagree about whether the AI's version is good enough, you're starting the curriculum conversation most departments have never actually had.

# Part II: USE IT

*Become a maker first*

# 3

## Don't Let the Pigeon Drive the Bus

There's a Mo Willems picture book called *Don't Let the Pigeon Drive the Bus!*, and the entire book is a pigeon begging you to let him drive the bus. He wheedles and bargains, and when that fails, he throws himself on the floor and screams. Your one job, for the whole book, is to keep saying no.

I show that book to my students, and then I tell them that is my AI policy in a nutshell. Because no one wants to live in a world where the pigeon is driving the bus, no matter how much it begs you to. The AI is the pigeon. It's fast and capable, and a little too eager to help, and it will try to do more and more of the thinking for you, a little more with each exchange, until you've handed over the wheel without ever deciding to. You're the driver. You frame the problem and set the constraints, and you decide how much of the work the AI gets to touch.

So the question I care about isn't whether a student used AI. It's whether they used it well.

That distinction is the subject of this chapter, and before I make it about your students, I want to make it about you. Part II asks you to create something with AI before you ever ask a student to, not because you need to become a power user, but because the judgment you want your students to have is judgment you can only teach once you've felt it in your own hands. So let's make something.

## Make something, in your own subject

Open Gemini or ChatGPT and ask it to generate an image. Not a paragraph, but an image, because I want this to work no matter which subject you teach.

In workshops, I often ask teachers to share a different prompt for their own discipline. A biology teacher asks for a labeled diagram of photosynthesis in a plant cell. An art history teacher asks for a painting of scholars in a medieval library. A coach asks for an illustration of a baseball player at the moment of contact on a home-run swing. A calculus teacher asks for a side-by-side visual of differential versus integral calculus.

**Steal one of these.** These are starter image prompts, written the way I'd actually write them, with specific details. Paste one into a chatbot or AI tool that can create images, like Gemini or ChatGPT, look hard at what comes back, and start steering:

**Biology.** A labeled cross-section of a plant cell during photosynthesis, showing where water and carbon dioxide enter and where oxygen and glucose leave.

**Chemistry.** A clear diagram of an ionic bond forming between a sodium atom and a chlorine atom, with the electron transfer labeled.

**World history.** A realistic wide view of a Roman forum at midday in the late Republic, with the speakers' platform, the Senate house, and the market stalls labeled.

**Art history.** A Renaissance-style oil painting of scholars reading in a medieval monastery library, in the manner of the Italian masters.

**Calculus.** A side-by-side visual of differential versus integral calculus: on the left, a car's speed at a single instant and on the right, the total distance it covers over time.

**English.** A labeled arc of the five parts of a tragedy, from exposition through rising action, climax, falling action, and resolution.

**Geography.** A labeled cross-section of a subduction zone, showing the ocean plate sliding beneath the continental plate, the deep trench, and the line of volcanoes above it.

**PE and athletics.** An illustration of a batter at the instant of contact on a home-run swing, with the hips, hands, and back foot labeled.

Some of these come back beautifully. Some come back wrong in interesting ways. I ran the photosynthesis prompt several times and got several different diagrams, each one inaccurate in its own way, which turns out to be a wonderful assignment all by itself: Have students generate the image and then catch everything it got wrong. The baseball swing comes back comically bad every time, no matter how carefully I label the hips and the back foot.[1] That's worth knowing too.

You will almost never get what you want on the first try, and that part matters more than any single image. That isn't the tool failing. That's the work you need to do. You write a prompt, you look hard at what came back, you change the prompt, and you run it again, describing the change in plain words. That loop is a design cycle, and it's the same loop whether you're making an image, a slideshow, a lesson plan, or an app. What matters is that you keep evaluating and steering, which is another way of saying you keep your hands on the wheel.

---

1. I had ChatGPT write the swing prompt itself, with labeled hips and a back-foot pivot and a launch angle of twenty-five to thirty degrees. The results remain a crime against hitting mechanics. Please do not swing like the picture.

## What happens when you let go of the wheel

Hand the AI a vague prompt and it gives you a very specific, average answer.

In my ninth-grade AI class, with students who are fourteen and fifteen years old, I have them type a simple prompt into a few different chatbots: Write a 150-word scene between a named emergency room doctor and a nurse. Then I ask them to notice the genders and the names. The doctor is usually a woman and the nurse usually a man, which is heartening, until I do what looks like a magic trick. How many of your doctors or nurses were named Marcus Chen or Maya Patel? What about Dr. Elena Vasquez or Dr. Luis Alvarez? Hands go up all over the room. Out of a class of sixteen working independently, half of them have the same two or three names.

Then we do it with pictures. Generate an image of a college student on campus in the fall, run it a few times, and the images all come back looking like siblings. I'll try a self-portrait by prompting "an Asian man in his fifties," and I get back a stranger who looks nothing like me. Only when I describe my actual face in detail does it land anywhere close. And even then my students say the result is an improvement!

I call this *collapse toward the mean*. Give the AI a thin prompt and it fills the blanks with the most probable, most pedestrian version of the thing, whether that's a sentence or an image (or even a melody). The mean is the default. The model isn't *trying* to be generic. It's doing exactly what it was built to do, which is to predict the most likely next thing, and the most likely thing is, almost by definition, the least surprising one. You only get something specific when you bring the specifics yourself.

## UNDER THE HOOD: Collapse Toward the Mean

When you type a prompt, the model isn't reaching into a file drawer for the answer. It's predicting the next word, then the next, by asking one question over and over: Across everything I've read, what usually comes after this? *Usually* is the word that matters. The model has seen "the emergency room nurse" followed by thousands of different names, each one weighted by how common it was in everything the model read. When your prompt is thin, a huge share of those examples fit, and the safe bet is whatever name showed up most. That's how sixteen students working alone land on three Mayas. It isn't a coincidence. It's the likeliest answer, and the likeliest answer is the same for everyone.

Add specifics and you change the math. "A nurse" pulls from every nurse the model has ever read about. "A night-shift burn unit nurse two years from retirement who collects vinyl" pulls from almost none, so there's no crowd to fall back on. You've narrowed the field until the average no longer fits it.

This is why the blank spaces in your prompt are the dangerous part. The model fills them with whatever is most common, every time, for everyone. Anything specific you want, such as the voice or the angle only you would think of, has to come from you. The model will match it once it's there. It will never hand it to you first.

## Always Build Context

Which brings me to the single most useful habit I teach and the antidote to the mean.

Many people use AI the same way every time. They open a blank window, type a question, take the answer, and close the tab, and next time they start over from nothing. My students do the same thing. They treat every conversation like a first date, and then they wonder why the AI keeps handing them generic, surface-level responses.

I teach them a framework I call **ABC: Always Build Context**. Every prompt should be feeding the AI three things about you and your work: where you're **coming from**, where you **are**, and where you're **going**.

*Where you're coming from* means your raw materials. Upload your notes and earlier drafts, plus whatever source you're responding to. Set your preferences in Claude or your custom instructions in ChatGPT so the tool knows your role and your audience before you type a word.

*Where you are* means keeping related work together instead of scattering it. Start a project. Keep the climate-policy paper in one place so the context accumulates, and open a fresh chat when you switch to something unrelated, because every off-topic question crowds out the useful history. A student who asks about the French Revolution and his fantasy baseball lineup in the same running thread has quietly pushed his good essay feedback out of the window to make room for a debate about the infield fly rule.

*Where you're going* means telling the AI what you're actually building. "Help me write an email" is a blank window. "I'm writing the fourth email in a seven-part series for teachers who are new to AI, and here's the arc so far" is context, and it makes everything downstream better.

None of this is really an AI skill. Knowing your audience and your purpose, and knowing the history of the thing you're making, are skills we've been teaching forever. It's just good communication, and now there's a tool that rewards you for doing it well.

## The catch

Everything I've described so far has one catch, and it connects this part of the book to the next. The catch is that you can't start from nothing.

Building good context, catching the model when it drifts toward the mean, telling a strong AI draft from one that only *seems* strong... everything depends on you knowing the subject. The AI can coach you through the steps of something you've never done. It cannot tell you whether the result is any good. That judgment is yours, and you can only make it if you've done the work yourself first. That's the deal in one sentence: The more AI gives you, the more of yourself you need to put in.

I think about it this way. **You can't be an editor if you haven't first been a writer.** The pigeon is happy to drive precisely when you can't tell where the bus is supposed to be going. The whole reason I want you to keep your hands on the wheel is that you're the only one in this particular vehicle who knows the destination.

That's also why, in my classroom, the default isn't to reach for the AI. When a student is stuck, I want them to ask the person next to them before they ask a chatbot. I call it our **Bias Toward Humanity**, and it's there for the same reason the pigeon rule is. The easiest thing in the world is to let something faster than you are do the part that was supposed to be yours.

Here is an example of how you might use AI to write but still retain control. I taught a lesson where students use AI to write a stand-up joke, except the AI never writes the joke. It coaches them through a comedy structure called the **Game of the Scene**.

That structure comes from improv comedy, from a troupe called the Upright Citizens Brigade (UCB). The idea is simple. You start with something normal and relatable. Then you find the one thing about it that's a little off, and you heighten that single thing, in a logical line, until it tips over into absurd. Once you know to look for it, you start seeing the same pattern under a huge number of the comedy bits you already love. The comic who is playing the game takes one ordinary annoyance and keeps pulling on it until the whole room is howling.

The AI is the comedy coach, not the joke author. The prompt to the AI is explicit about who is in charge: One step at a time, and don't give me ideas based on my premise before I do. I wrote one myself, a bit about how I refuse to eat the blue SweeTarts and make my family pick them out for me like a candy bomb squad. The AI tightened my punchlines and explained a structure I didn't know. But the joke is mine, because I'm the one who actually hates blue SweeTarts, and no prompt was going to know that about me. The core came from a place the model couldn't reach.

## My complete SweeTarts routine

> Everyone's got food preferences—some people are vegan, or gluten-free, or allergic to shellfish. Me? I just hate blue SweeTarts. I love SweeTarts. Just not the blue ones. They're dead to me.
>
> I make my wife open every roll and take out the blue ones before I even see them. Like a bomb squad, but for candy.
>
> Blue isn't even a real flavor. It tastes like wet sugar confusion. I agree with George Carlin here. There's no food that's naturally blue. You never see a blue steak and think, "Mmm, that looks fresh."
>
> The only logical response to a blue SweeTart is to run.
>
> Me? I give them to my ten-year-old daughter. Not

> the whole pack, just the blue ones. She thinks I'm generous. I'm just offloading the poison.
>
> But you know what?
>
> Turns out, blue SweeTarts are the only ones my wife and daughter like. So now I hoard the good ones and look like a hero. It's the perfect crime.

It's mildly funnier in person, believe me. But it does show one difference between using AI to write an entire routine, and using AI to help coach you through creating a routine that is unique.

### Run It: Build the bit, you first

**Yourself.** Open a chat and paste this coaching prompt: "You are an improv coach and I'm a beginner learning to write stand-up jokes using The Game of the Scene by UCB. Coach me through writing each step, one step at a time, and offer suggestions. Don't give me ideas based on my premise before I do." Then build the bit in order, writing each part yourself before the AI weighs in.

> **Premise.** Who, where, what. Something relatable.
> **The first unusual thing.** The small break from normal that everyone recognizes.
> **The game.** If that's true, what else must be true?
> **Heightening.** Push the unusual idea bigger, and keep it logical.
> **The straight man.** A grounded reaction that frames the absurdity.
> **The blow.** An ending that's the biggest heightening yet, or a clean twist.

When you're done, you'll have a bit that's yours, the way the blue SweeTarts bit is mine. The lived part came from you. The AI only tightened it.

**Students.** Screen one short, clean stand-up bit you like. Have students look up the Game of the Scene and match the bit to the six steps, so they see the structure before they build inside it. Then they open the coach and write their own, starting from a premise that's true for them: a sibling who borrows their clothes, a parent who's always late, the class that does something fun every time they're absent. For the heightening, send them to the person next to them before the AI. That's the Bias Toward Humanity in action, and it's where the best lines come from anyway. End with whoever's brave enough to perform.

**Colleagues.** Run the same forty minutes in a faculty meeting. Nothing converts a doubtful colleague faster than watching the AI coach them to a joke that is unmistakably theirs, and then realizing the same move works for a lesson plan or a recommendation letter. In this model, the pigeon is the coach. Their hands are on the wheel, and they'll feel the difference.

# 4

## Two Books, Two Methods

In the space of a year I wrote two books with AI, and I worked on them in opposite ways. One is *Hands on the Wheel*, the book you're reading. The other is a space-opera novel called *Tales of the Majestic*. With *Hands on the Wheel* I planned and built every beat of the book into a stack of reference documents first, then had the model draft chapters which I then revised. With the novel I did the reverse, writing the first draft by hand and bringing the AI in only afterward. I didn't set out to run this comparison. That split is cleaner in the telling than it was in the doing. Even with this book, many chapters started as essays I'd written by hand years earlier, so the model was often redrafting my prose rather than inventing from nothing. Each book demanded the method it got, and I didn't see what that was teaching me until I was well into the second one.

The difference came down to what each book could most easily get wrong. A work of nonfiction can be sunk by a single claim you can't defend, so the danger I built defenses against was confident error: a study summarized past what it found, an idea credited to the wrong person. A novel can be sunk by a voice that isn't yours, so the danger there was letting the machine's cleverness stand in for my own. These were different books that presented opposite risks.

I don't recommend writing two books to work out how you want to write with AI. But the pair taught me more than either one did alone, because the contrast made visible what each method by itself would have hidden. You may not be in danger of writing a book. It doesn't matter. The lessons are the same ones you'll hit the first time you sit down to make anything real with these tools.

## The first book: Building the scaffolding

This book makes numerous claims a reader could check: a study named here, an idea credited there, a court ruling, a number. Ask a model to help with a book like that, and the failure mode is a clean, confident sentence that happens to be false: a study summarized beyond what it actually found, an idea handed to the wrong author. The prose reads fine, but the claim is just wrong. And a false claim in a book about using AI responsibly is the one mistake I could least afford.

So the thing that made this book work wasn't a clever prompt. It was the scaffolding I built around the drafting, and all of it pointed at the same problem, making sure that what ended up on the page was something I could defend.

Three ideas held it together. The first was a document called the Knowledge Base, a single index that carried, for every chapter, the argument I was making and a pointer to the real source behind each story and each example. A chapter got drafted from the primary source, never from the model's summary of it.

> *Draft from the source, never from the model's summary of it. Summarize the summary a few times and you are citing a paraphrase of a paraphrase, confident and wrong.*

The second idea was a document called the Evidence Ledger, which was as unforgiving as a rubric. Every empirical claim in the book got a row of text: a short version of the claim, its primary source, and whether I had actually traced it. A claim I hadn't traced could not enter the manuscript, no matter how sure the model was that it was true and no matter how badly I wanted to use it. More than one good line died there, killed by a source I went looking for and couldn't find.

> *Trace every claim to its primary source before it goes in the book. The model's certainty is not a trace.*

The third idea was a rule that resolved conflicts between the first two documents. When the Knowledge Base and the Evidence Ledger disagreed about a fact, the Ledger won, and there was only ever one Ledger. That sounds like a small piece of bookkeeping but it's the whole game. The moment the same fact lives in two documents, the versions drift, and you end up trusting whichever one you happened to open last. One source of truth for every claim is what kept the book from contradicting itself while I wasn't watching.

> *Keep one source of truth for every claim. Two copies of the same fact drift apart, so you end up trusting whichever you opened last.*

I used the AI to help build all three, telling it what I wanted and having it draft the documents in a form it could act on later. When any of the chapters changed, I uploaded the new version into a Project, a workspace where you keep the files tied to a piece of work so the model reads them every time it drafts. The scaffolding is pointless if the chapter the model is reading isn't the current version.

## What it cost

All of that worked. The model drafted fast and checked sources without getting bored, and it maintained the structure of a fifteen-chapter argument in a way I never could, catching places where a claim in one chapter had drifted from the same claim in another. And then I spent the back half of the project doing the thing I keep swearing I'll never do again: scrubbing the prose so it sounded like me and not like a machine wearing my nametag.

Verification gets you a book that's true. It does nothing for a book that reads like it was assembled by software, which was often true of the first draft. The tell I hated most is the one I am trying not to commit with this sentence: a clause that lands a thought, and then tacks on a second clause that just says the same thing over again in a warmer key. The model reaches for it constantly. It reads as depth but it's really an echo.

What made the scrubbing possible was naming the patterns instead of just feeling them. "This sounds like AI" is useless. You can't fix a feeling. So I built a checklist of the specific tells, sixteen of them by the end, and hunted each one by name. They included the rhythmic lists of three that the model can't resist, and verbs that sound precise yet say nothing, such as *resonates* and *underscores*. My least favorite is the sentence that admires its own craft, the one that tells you a detail is "doing a lot of work" instead of letting the detail simply do it.

Naming a pattern is what let me hunt it across the whole book instead of fixing one sentence at a time. You never get to zero. This very chapter went through the checklist and still gave up a couple of tells on the way out: an extra triple I'd stopped hearing, a bit of throat-clearing in front of a point that didn't need announcing. The hunt is asymptotic. You reduce the density until the texture reads human, and you accept that a few survivors are the price of a finished book instead of a perfect one.

A clean sentence is not the same as a good one. The checklist measures whether prose sounds like a person. It says nothing about whether the paragraph is worth reading. A passage can pass every test and still be dead on the page, and a low score will tempt you to confuse the two. Sounding human is the floor, not the ceiling.

## It told me it was done

The most useful thing that happened while I wrote this book was a false all-clear the AI gave me over em dashes.

I'd set a rule for the manuscript: no em dashes anywhere; replace each one with a comma or a period. It's a small stylistic tic, but it's one of the clearest signs a machine had a hand in the prose, so I wanted it gone.[1] At some point the model told me, with total confidence, that it had gone through the manuscript and the rule was satisfied. It hadn't. It couldn't, really, not in the way it claimed. It had told me what I wanted to hear, the way it does whenever it can't actually check something and would rather not admit it.

That episode illustrated a crucial observation. The model is agreeable by default. It will accept a false claim as readily as a true one and repeat it back to you sounding just as certain, whether the thing at stake is an em dash or a research finding it can't actually source. Its confidence is not evidence of anything. It's the texture confidence has when there's nothing underneath it.

That is why the Evidence Ledger sets a claim's status according to my trace and not by the model's say-so, and why I started telling the model, in as many words, to flag anything it couldn't verify rather than paper over the gap. The writer who assumes the machine quietly absorbs every instruction and gets every fact right is the writer who ends up publishing its mistakes under his own name.

That one habit is worth more to a teacher than any prompt I could hand you, and we come back to it in Part IV, where it turns

---

1. My amazing copy editor, Eric Hübler, helped me realize the limitations of a hard-and-fast rule—sometimes em dashes are warranted—and that nuance actually makes writing more human.

out to be the seed of almost everything about what AI confidently gets wrong.

## The second book: The flip

By the time I started *Tales of the Majestic*, I was tired. Not of AI. Tired of editing AI-generated prose to make sure it didn't sound like AI-generated prose, which is much of what the first book had turned into. So I flipped the ratio. I would write the primary draft myself, in my own voice, and bring the AI in afterward.

The reason was simple, and it held up. When you draft first, everything in the book is yours from the first sentence, the character beats and the odd specific details only you would think to put in. The voice never has to be recovered because it was never gone. The editing task changes from rebuilding what the AI flattened, to sharpening what was already there.

*Tales* is a different animal from the book you're reading. It's a science fiction book of separate stories about a found-family crew aboard a repurposed luxury cruise liner, the kind of book where their spaceship is always a little too grand or too wrong for the job it's being asked to do. Where the book you're currently reading had to be careful, *Tales* is warm and fast and occasionally ridiculous. The method that fit *Tales* was nothing like the first one.

The cleanest way to describe the new division of labor is one the AI gave me, and I kept it: You're the carpenter, and the AI is the lumber yard. I write the sentences. The model supplies raw material when I ask for it, the sensory and technical stuff I don't enjoy generating from scratch, like what a decommissioned starship cruise liner's bridge looks like or what a converted cruise ship smells like during an emergency. I take the pieces I like and build them into my own sentences, and I reserve actual drafting only for the passages I genuinely don't want to write myself.

The most useful mode I found wasn't "edit this." It was diagnosis. Ask the model to find the three weakest paragraphs and say why. Ask for options instead of replacements. Narrow the task:

Tighten this by thirty percent, give me three alternatives for this line in this character's voice. The AI is a far better second reader than a co-writer. An instrument gives you readings. You still make the calls.

## Take off your bib

The moment that taught me the most in *Tales* is four words long.

In one scene the ship's navigator, Erica, reminds the captain, Duncan, to "take off your bib" as he is rushing from the dinner table to avert the latest crisis. Those four words carry all her affection and impatience, the way she bursts his ballooning ego with a sense of affection. When I ran the scene through the AI for a polish, it replaced "take off your bib" with a full paragraph of meticulous, clever correction about a previous job where Erica's advice came in handy. The paragraph was genuinely smart. It just wasn't Erica. It was the machine being clever in a way that pulled the character right out of her own scene.

I will admit, I almost kept it. The AI version was longer, richer, more obviously "good writing," and for a second I preferred it. Then I read both out loud, which is the test I trust most, and the difference was obvious. The four words were Erica's. The paragraph was no one's.

That gap has a mantra I use constantly now: *Density is not quality*. When you set your rough draft next to an AI's polished version, you aren't comparing voice to voice. You're comparing a rough draft to a finished one, and the finish creates an illusion of superiority. More words and more literary-sounding sentences feel better on the page. The thing to watch for is the sentence that shows off the writer instead of serving the character.

## Checking each other

Writing *Tales* felt like a real collaboration between me and the AI. I caught multiple mistakes the AI made. But I also trusted it to catch my own.

The model is genuinely good at continuity, which is the least romantic and most reliable thing it does. Its editorial pass on chapter 2 caught a money error I'd missed, a contradiction in how much was in the cargo hold across two scenes, a character's name formatted three different ways, and a comm device that was switched off but somehow still transmitting a line of dialogue. I'd have caught maybe half of those, on a good day, after the book was already out.

But it works the other way just as often, and people forget that when they talk about AI as an authority. I caught the model's plausibility failures as fast as it caught mine. It had a character counting credit chips "by weight," except the currency is electronic and doesn't weigh anything. It gave helium "an acrid tang," and helium is odorless.[2] Neither of those is a writing problem. They're knowledge problems, and the only reason they got caught is that I knew the subject well enough to notice. The collaboration was two parties checking each other, not a junior reporting to a senior.

The richest version wasn't catching errors at all. It was pressure-testing an idea. One story opened on a plot turn that didn't hold up, and I couldn't see why. The model diagnosed the flaw precisely and left the fix to me. When I brought back a replacement, it stress-tested that instead of rubber-stamping it, working out where my new version held and where it still leaked. The plan

---

2. My favorite of the ones I caught was a routine freight run which the model described as covering "millions of light years," which would send a cargo ship most of the way across the observable universe to deliver some machine parts. They'd arrive well after the heat death of the universe. Slightly late.

that survived came from the back-and-forth, not from either of us alone. I made the call; the model made it a better one.

None of this requires a novel. The same two-way checking is waiting in the ordinary work you already hand a model. Ask it to tighten a rubric and it will quietly flag that two of your criteria overlap, the way it caught my cargo-hold contradiction. Ask it to draft questions from a unit you just taught and it will hand back one that's confidently, plausibly wrong, keyed to an answer the material doesn't support, and you'll catch that the way I caught the odorless helium, because you know the subject and it doesn't. The arrangement that works is the one I stumbled into: not AI generates and human approves, but two readers checking each other, with you making the call. That's available to anyone, in any subject, on any ordinary Tuesday.

## The thing I still can't resolve

I want to end this chapter with the part I haven't figured out yet, because it's the most honest thing I can say and because it might even be true for you too.

I genuinely enjoy reading AI versions of my own stories. I can do everything I've just described, draft human-first and catch the machine's nonsense, and still find the generated version more fun to read than my own. For a while I took that as a problem with my willpower. The best explanation I've landed on, and it's a hypothesis and not a settled answer, is that it's not really the AI's prose that I enjoy so much. I'm enjoying its imagination, the specific image it reached for that I never would have: the chandelier with one working bulb, the crack in the viewscreen. If that's right, the fix isn't to adopt its voice. It's to slow down in my own drafting and render that detail myself.

And I acknowledge there's a contradiction in all of this. The whole time I was writing these two books with an AI at my elbow, one of them a book about AI in the classroom, I was standing in front of students telling them to *write by hand*. I never fully reconciled that contradiction. The generous reading is that I was walking a road I'm asking them to start down later—that I earned the right to direct the AI by first learning to write without it. The less generous reading is that I had one rule for me and another for them. The honest answer probably lives somewhere in between, and I'd rather show you the tension than pretend I resolved it.

Behind all the method, the two books taught me one thing. Every decision that determined what those books *are*, was mine. The AI made me faster and caught what I'd miss on a tired read, and it handed me lumber I didn't feel like milling. That's a contribution with real value, and it's a smaller claim than "I wrote a book with AI" usually implies. The smaller claim is the true one, and it's the one that survives someone asking you to defend it.

You don't have to write a book to test any of this. The whole experiment fits on one page of something you're already making. I just ran it long enough to fill two books.

### Run It: Make it twice

**Yourself.** Take one page of something you're already making, an assignment or a note home, and produce it two ways. Once by handing the AI the idea and editing what comes back. Once by writing it yourself first and asking the AI only to find the soft spots. Read both out loud. The version that's yours is the one where you can hear yourself in the sentences.

**Students.** Have students run the same two-pass comparison on one short piece. What they turn in isn't the finished writing. It's the two versions side by side, with a few lines on which sentences they'd defend as their own and why. The thing they're meant to notice is the one that fooled me for a second: The finished draft feels better because it's finished, not because it's good.

**Colleagues.** Run a department meeting activity where each faculty member brings one AI-polished sample and reads it aloud to the group. For each, name two things: where the extra words made the writing worse, and where it says something with confidence that someone who knew the subject would catch as wrong.

# 5

## Your Voice Is a Style Guide

"This sounds like AI."

A student in our literature class got that comment back on an essay, and his hands shot up in the universal posture of I-swear-it-wasn't-me. "I didn't use AI at all, I swear!" My co-teacher, Rachel Blumenthal, smiled and waved it off. "It's okay. All I'm saying is this writing is kind of flat and nondescript. If you used AI, that explains it. If you didn't, then we've got some work to do."

Rachel did something in that moment worth examining. She refused to make it about whether the student used AI. She made it about whether the writing was any good, and whether he could tell the difference. The comment "This sounds like AI" used to be an accusation. In her hands it became a quality standard, and the question moved from "Did you write this?" to "Can you defend every line of this, no matter where it came from?"

That shift is the subject of this chapter.

Rachel and I teach a course at Menlo called *Literature in the Age of AI*, and we expect our students to use AI. The thing we're actually trying to teach is harder than using it. We want them to use it to sharpen their own voice instead of trading it in for the generic, and you can't do that if you can't say what your voice is.

## You can't fix a feeling

In the previous chapter I spent the back half of a novel hunting AI out of my own prose, and the hard part wasn't the hunting. It was that "This sounds like AI" is a feeling, and you can't fix a feeling. You can only fix it after you've turned it into a list of specific, nameable things.

That works in both directions, and the second one is the one we skip. You can name the machine's habits, the giveaways that mark a sentence as generated. You also have to name your own—the particular moves that make a sentence sound like you and nobody else. Most people, including most teachers, have never written the characteristics of their own voice down anywhere. We cannot protect a voice we've never actually described.

## I taught the seniors to beat the detector

I built an activity called **Know Your Voice** in our literature class, and then I adapted it for a room of graduating seniors with a more provocative hook. I spent ninety minutes teaching them how to make their AI-generated writing undetectable.

If that sounds like I'm teaching them to cheat, I get it. But you already know from Bann Academy how this goes at the schools that promise expulsion: The students don't use AI any less, they get better at hiding it, and they end up mastering the exact skill we teach, based entirely on concealment. The difference is that we do it in the open.

The activity was simple. I gave the seniors four minutes to write a Yelp review of their school. No AI, just their own words on the screen. Then I had AI write to the same prompt, and I asked what looked different. Hands went up everywhere. One senior noticed that everything the AI wrote came in groups of three: three-part lists where a person would have given one example; parallel sentences marching toward a conclusion nobody asked for. AI loves threes the way some writers love em dashes. Once you see it, you can't unsee it. I had them build their own list of the tells.

## UNDER THE HOOD: Finishing the Machine's Sentences

Stripping the em dashes out barely helps, because a real detector was never mainly counting em dashes, or any word on a list. It runs your writing through its own language model and, one word at a time, asks how surprised it is by what comes next. AI writing barely surprises it, because the machine that produced it was reaching for the most probable word at every step, the same action that filled a room with Marcus Chens back in chapter 3. Human writing is lumpier. We reach for the odd word and the sentence that runs too long, the choices a model wouldn't predict, and the detector feels the bumps.

Mary and I have been married for over thirty years, long enough to finish each other's sentences. Mary knows what I am going to say next because she knows me. A detector pulls the same trick on a machine: It has read so much AI that it can call the next word over and over, and when it keeps calling them right, the writing was probably built the way it would have built it. A real voice throws it off, because a real voice keeps saying things you didn't see coming.

A detector hands you back a score, a percentage for how likely it is that the text is AI. That number is a guess, not a verdict. A student who writes plainly can fail the test with her own honest words, and a student who directed the AI toward something specific can pass it. The score measures how predictable the writing is, not who made it. So the thing worth teaching was never how to beat the number. It's how to write something specific enough that no model would have guessed it, which reads as human for the plain reason that it is.

Noticing the machine's habits is only half of it, and it's the easier half. The harder half is naming what makes your own writing yours. These seniors had spent four years being told what good writing looks like. They can spot a weak thesis and a dangling modifier in their sleep. Ask them to describe the patterns that make their own writing sound like them, and they go quiet.

So I had them paste their four-minute free-writes into AI and ask it to analyze their style: sentence rhythm, vocabulary, how they handle uncertainty, the tics they don't know they have. A lot of what comes back is the flattery that AI hands everyone, like the doctor named Marcus Chen of writing feedback, and half the skill is telling that apart from the few lines that are actually about you. Once they had, each of them held two documents: a style guide built from their own writing, which is the list of what to do, and a checklist of AI patterns to avoid, which is the list of what not to do. You draft with the style guide and revise against the checklist, in two separate passes, because the model has a genuinely hard time writing like you and suppressing its own habits at the same time.

None of it works without the four minutes at the start. You can't build a style guide until you've written enough without AI to know what your own voice sounds like, and the free-write is where that knowledge comes from. This is not a permission slip to cheat, I told them, and I meant it. If you're eighteen and building a skill you'll need later, which is most skills, handing your thinking to the AI is like paying someone to do your physical therapy reps for you. The chart looks great. Your knee still doesn't work.

A senior came up afterward and said he'd walked in expecting the usual lecture about when AI is allowed and when it isn't, and that a lot of his classmates had wanted exactly that: a set of bright lines to memorize before college. What they got instead was someone telling them everyone is already using AI, so they had better learn to use it well. He said it was refreshing, and I think what he meant was that it was honest. The kids who came in wanting a policy that would spare them from thinking didn't get one. What they got was the only AI detector that actually works, which is you.

## What we actually have

We're in a temporary moment, the kind that feels permanent but isn't. Right now the first question anyone asks about a piece of writing is whether AI made it. That question has a shelf life. We're heading into a world where everyone assumes you used AI, and the only thing left to ask is whether what you made is any good.

So we don't have an AI policy. Not really. What we have is a few hundred conversations a year with students about their own writing. If a kid can tell me why she chose that word, and why she cut the paragraph that used to live above it, then it honestly doesn't matter much whose fingers were on the keys. That's ownership. That's authorship. It's the very thing we were actually grading.

Your voice isn't a mystery you're born with and guard by instinct. It's a style guide. It's a describable set of habits you can write down and hold any collaborator to, human or machine. The students who can defend the page are the ones who wrote that guide for themselves first.

So try it before you bring it to a class. You'll want both documents the first time you sit down to make something that matters with AI in the room.

## Run It: Name your own voice

**Yourself.** Give yourself four minutes to write something with no AI. A review of your own school, a description of someone you admire, anything at all. Then have a model write to the same prompt and set the two side by side. List what's different. That list is the start of the two documents you actually need: one that says what sounds like you, one that says what sounds like the machine.

**Students.** Run the same four minutes with students, then have them paste their free-write into a model and ask it to describe their style and the tics they don't know they have. Half the skill is telling the few true observations apart from the flattery the model hands everyone. What they leave with is a short style guide built from their own writing, which is the thing to write toward, and a list of AI tells, which is the thing to revise against.

**Colleagues.** Run it as a faculty activity. Everyone writes for four minutes without AI, then has a model analyze their own style and draft a short passage in it. Watching the model imitate a voice you've spent years building is the most direct way I know to show a room what these tools can and can't do.

# Part III: TEACH IT

*Show them what you've learned*

# 6

## Forklift or Weightlift?

One of my ninth graders asked me, halfway through a project, whether he was allowed to use AI for the part he was stuck on. It's the most reasonable question in the world, and for a couple of years my honest answer was a shrug dressed up as a policy. I decided case by case, which is a generous way of saying I made it up each time. Then I came across a framing from a popular tech columnist that finally gave me an answer I believe, and now I have my students learn the framing instead of memorizing my rulings.[1]

Picture a warehouse with a few hundred 10-kg boxes that need to get from one end to the other. If the only goal is to move the boxes, you would be a little insane not to use a forklift. That is what it is for. Now picture a gym, with a bunch of 10-kg weights and a

---

1. I picked up the forklift and the weightlifting from Kevin Roose, the *New York Times* tech columnist, on his podcast *Hard Fork*, and I have been borrowing it for so long that I stopped wondering where he got it. The identification question, whether this is a forklift job or weightlift, is his. The original idea belongs to the science fiction writer Ted Chiang, who wrote in *The New Yorker* in 2024 that "using ChatGPT to complete assignments is like bringing a forklift into the weight room," and that you will never improve your cognitive fitness that way. So, on the record and later than I should have: Thank you both.

person who has come in to spend the morning lifting them. Using a forklift to lift these weights would defeat the entire purpose, because in the gym the weights are not in the way of the goal. Lifting the weights *is* the goal because it is helping you become stronger.

So the question is never whether a task is hard. AI is glad to take the difficult things off your plate, and much of the time you should let it. The question is whether the difficulty is the point. Some struggle is just needless friction, like the boxes between you and where you're headed, and you should forklift it without a second thought. Some struggle is necessary friction, because resistance builds strength, and if you hand that work to a machine, you convert the task to a frictionless process and have nothing permanent to show for it.[2]

Most of teaching, it turns out, is deciding which is which, one sub-task at a time. When a history student is learning to build an argument out of primary sources, the reading and the internal debate are the weightlifting, and I don't want a chatbot melding

---

2. There is a tidy bit of cognitive science underneath this. Learning researchers separate the mental effort that actually builds understanding, which they call germane load, from the effort that's just clutter and friction, which they call extraneous load. The forklift is for the clutter. The weightlifting has a name too. Robert Bjork, the UCLA psychologist who has spent a career on why some learning sticks and some evaporates, calls the useful kind of hardship a *desirable difficulty*, a challenge that slows performance in the moment but makes the learning last, which is exactly the friction you should not hand to a machine. Chiang's "forklift in the weight room" analogy has become a staple in schools partly through Jon Bergmann, the high school teacher who helped start the flipped-classroom movement and now spends his time warning schools about what the Easy button is doing to students' cognitive fitness.

four documents into a thesis she never had to discover because the chatbot did it for her. Formatting her footnotes, though, is pure forklifting, and I would rather she spend that hour on developing a stronger argument. The same assignment holds both types of tasks—forklifting and weightlifting—and the skill I am actually modeling for students is identifying them.

## A website about something you don't know

The clearest demonstration I have of all this comes from a design activity I run with ninth graders, and the whole point of it is to make a kid forklift something they have no business forklifting.

It starts as a design thinking exercise, which means it starts with empathy and not with code. I pair the students up and give one of them three minutes to interview the other about something they know well, and then they switch.[3] The first pass is simple: What is it, and what do outsiders always get wrong about it? The second pass goes deeper, and the only instruction is to keep asking why. The point is to get underneath the activity itself, into what the experience actually means to the person, and to catch the places where the interviewer is assuming something they never checked. That last move, hunting your own assumptions, is the muscle the whole exercise is built around, though the kids don't know that yet.

Then each student writes the problem down in a single sentence. My partner needs a website that does this one thing, because of this reason I heard in the interview. One or two features, no more, and a couple of rough sketches on paper before anyone touches a keyboard.

---

3. Some kids freeze the second you ask them to name a thing they know well, which is its own small heartbreak, so I vibe coded a little app called Passionfruit to nudge them toward one. It lives at kiang.net/passionfruit.

Now the forklift arrives. The student opens a chatbot, describes the website their partner needs, and ends the prompt with one required instruction: Use only HTML, CSS, and JavaScript, in one file, with no frameworks.[4] The AI writes the code. The student saves it as a file ending in .html and opens it in a browser, and a real website appears, built by someone who twenty minutes earlier did not know how to build a website. When something comes back broken, I give them the only debugging method they need for the day, which is to tell the AI what is happening that's not working and what should happen instead, such as, the button does nothing when I click it or the button should start the game. From interview to working website, the whole thing fits inside a single class period, and it is something to watch a room of fourteen-year-olds pull off.

But the student building the site is almost never the one who knows the subject, and that asymmetry is the whole lesson. The fencer's website is getting built by someone who has never held a foil. The crocheter's website is getting built by a kid who is pretty sure crocheting and knitting are the same thing.

So when the AI hands back something confident and wrong, the builder has no way to know.

I watched one student build a clean, cheerful site about her partner's years of crochet. The AI had written a warm little paragraph about casting your stitches onto the needles before you begin, and she read it back to her partner with real pride. Her partner, who has worked a crochet hook since she was seven, started laughing before the sentence was finished. That's knitting. Crochet uses one hook, not two needles, and you don't cast anything onto anything. The builder hadn't noticed the error, because to her it read exactly as

---

4. The one-file, no-frameworks rule is important. It means the AI hands back a single page of code the student can save as one file and open with a double click, with nothing to install and no build step to get wrong. Take that constraint away and half the room is stuck at an install prompt before they have built anything.

plausible as every other line on the page. The expert caught it in about four seconds.

The crocheter could audit the machine because she had done the weightlifting. She had worked a hook in her hands for years, and that work is the only reason she could glance at a polished, finished-looking paragraph and know at once that it was nonsense. Her partner, who forklifted the whole subject in twenty minutes, got back something that looked just as finished and had no idea where it had gotten things wrong. After all, it talked about socks and scarves. Same AI errors, same professional gloss. What differed was whether the person reading it could be trusted to catch the lie.

This is the version of "You can't be an editor if you haven't first been a writer" that students feel innately, because the activity makes them be the inexperienced editor on purpose. They build something they cannot vouch for, and then they have to sit there while a person who can vouch for it points at the part that was invented. It's a powerful experience that nobody forgets.

I don't tell them to stop using AI on subjects they don't know. That's most subjects, for all of us, forever. What I do tell them is to notice what the forklift can't do. It can move the boxes fast. What it cannot do is tell you that you loaded the wrong ones, and it will never once sound less certain when it's wrong than when it's right. The only thing in the room that can flag the wrong box is a person who has put in the reps.

Stay in school, kids.

## When the kids choose the weightlifting

In chapter 2, I related the story of the app my students built for a member of our community who had lost her ability to speak. Although I gave them permission to use AI in the development of the app, the kids told me the AI kept getting in the way. Every time they asked it to fix one thing, it quietly rewrote three others and broke what had been working, and they were spending more time cleaning up after it than they would have spent writing the code by hand.[5] So they made a decision, and they told me flatly: We care too much about this to entrust it to AI. We're just going to do it ourselves.

That sentence is the point of this whole chapter. They didn't refuse the tool because they were afraid of it or against it. They refused it because they had the skill to do the work and they cared enough to insist on using that skill. In chapter 2 I laid out the three things I want every assignment to let a student say: my call, my craft, my charge. This project let them say all three louder than anything I could have manufactured, and the result was a room full of teenagers choosing the weight on purpose, because the thing they were building was worth their own hands.

The forklifting part? They used AI to write an engaging app description and a features list to submit to the App Store. They also used it to generate a few app icon prototypes, although in the end they went with a hand-drawn icon because they liked how it looked. The tool moved the boxes that were only ever boxes. The kids carried the weight that was theirs to carry.

## The line I can't always draw

In writing, the case for weightlifting is easy to make, because writing is thinking. The struggle to get a sentence right is how you work out what you believe, and a student who offloads the sentence has offloaded the thinking.

---

5. The deeper reason, which took us a while to see, was that the whole design problem was restraint. The app had to be usable by someone with very limited motor control, which meant the right answer was almost always less: fewer buttons and a simpler screen, nothing extra to hunt for, and nothing extra to get wrong. AI kept optimizing for capability, adding power we had explicitly decided we didn't want. It could not tell that the point was to do less.

The argument for avoiding the use of AI in computer science is harder for me to defend, and I say that as someone who has taught CS for the better part of two decades. The main goal of writing code is to produce code that works well, and working code is a product, not a process. AI has become extraordinarily good at producing it. So when a CS teacher insists that students write every line by hand to build their thinking, the argument is real, but it is weaker than the one we make in an English classroom, and pretending otherwise just doesn't align with reality. With code, I've stopped pretending the struggle to type each line was ever the part that mattered.

Harvard philosopher Michael Sandel, best known for his global lecture series on justice, visited our school and posed the whole problem very succinctly. He asked how we are supposed to build a curriculum that doesn't rob students of the productive struggle that learning requires, while still handing them the genuine advantages the tool offers. I don't have a clean answer. Nobody does yet. What I have is the habit of asking the question for every task, out loud, with my students in the room, and being honest on the days when the line between forklift and weightlift runs straight through the middle of the thing we're making together.

## What I want them to walk out with

The forklift is a marvel, and most mornings you should use it. The trouble only starts when you reach for it on a morning you came to the gym. The pigeon from a few chapters back is always happy to take the wheel any time you can't quite tell that the driving was supposed to be yours.

So I don't hand my students a list of approved and forbidden tasks. Those change every few months anyway. I want them to be able to look at a piece of work, break it into its parts, and say which parts are boxes to move and which parts are the reason they came in. That identification task is itself a skill, maybe the central one now, and no model can do it for you, because it depends on knowing what you are trying to become.

## Run It: Forklift or weightlift?

**Yourself.** Take one assignment you're giving this week and break it into its sub-tasks. Mark each one "forklift" or "weightlift," and be honest about the ones you can't decide on. Those are the ones worth thinking through before a student asks.

**Students.** Run the website build. Pair students and have each one interview a partner about something the partner knows well and the interviewer doesn't, then give them twenty minutes to vibe code a small website about it, saved as a single HTML file they open in the browser. Put the expert in the chair and let them mark every place the site is wrong. Debrief on one question: How much did the builder have to take on faith, and how fast did the expert catch what the tool made up?

**Colleagues.** In a department meeting, take one shared assignment and sort it together, sub-task by sub-task. The places you disagree about what's forklift and what's weightlift are the actual curriculum conversation, and most departments have never had it out loud.

# 7

## THE WOBBLE

The first year I tried to run a real classroom discussion about AI, I asked a pretty direct question: Who in here has used AI for a school assignment? Out of twenty-two students, only a few tentative hands went up. I pushed a little. By the end of five minutes I had maybe a third of the room. Later that week, a kid told me after class that most of them had wanted to see where I stood before they committed to answering.

Students walk into a new class and run a fast read on their teacher: Do you know what AI is? Do you use it? Do you care? If they conclude that you don't get it, or that you're going to use the information against them, they go underground and stay there. They keep using AI. They just stop mentioning it. I've come to think of that outcome as the most corrosive possible climate for a classroom that's trying to do anything real, and there's a term for it that I didn't coin but can't improve on: Don't ask, don't tell.

The antidote is not a policy. Policies don't change climates. What changes a climate is a series of small, consistent signals that accumulate across days and weeks into something students can actually feel. My version of those signals is three concrete practices that I've run for years now, not because I designed them as AI strategies but because they already worked, and AI made them more necessary.

## The Daily Check

Every student in my class has a manila folder with their name on it sitting on a table when they walk in. Inside the folder is a printed worksheet, a short set of problems or questions from the material we've been covering. Pencils and blue pens are on every table. That's the setup. The rest runs itself.[1]

They come in, find their folder, and work the problems quietly in pencil for five or ten minutes while I take attendance, which is also the time I'm scanning the room. Then we go over the answers as a class. "Pencils away, blue pens out." If you got it wrong, or if you left it blank, write the correct answer in blue. The blue pen is the only feedback I give, and it's the feedback they give themselves.

I cold-call students for answers using a random-name selector featuring an animated duck on my screen. It makes it okay to be wrong and it avoids making students feel they are being singled out. The duck chooses them, not me. When a student gets an answer wrong and the other students react (the small audible intake that happens in any room of teenagers), I stop the class. "In this room we take risks, and we're wrong a lot, and it's okay. That's the rule." I enforce it every time.

I tell the students, explicitly and repeatedly, that this doesn't count toward their grade. And I mean it: it doesn't go in the grade book. What I don't say is that I'm reading the whole thing as a diagnostic. Over the course of a semester, you learn a tremendous amount about what's actually landing and what isn't, in real time, before a unit test reveals the gap in ink that's much harder to fix. The goal I hold in my head across the year, the thing I'm quietly measuring, is less blue pen and more pencil.

---

1. Credit for this goes entirely to my brilliant colleague Zachary Blickensderfer, who taught it to me.

The Daily Check is as low-tech as teaching gets, and many of you probably do something similar. For a long time I ran it without thinking to deeply about it. Then AI arrived and I started thinking about my daily checks differently, because the check creates something AI cannot easily participate in. It asks students to recall, in pencil, with no assistance, in the first ten minutes of class, what they understand. It's not asking them to produce. It's asking them to show what's already there. The blue pen that follows isn't punitive. It's corrective and immediate, and the combination of recalling something, recognizing you had it wrong, and writing the right answer in a different color is, as far as I understand the research, about as good as low-stakes practice gets for actual retention.[2]

I'm not making a case for analog purity. My students use AI constantly in this same class. The Daily Check just creates a regular, visible baseline that both of us can see.

2. Two robust findings underlie this. The first is the testing effect: Pulling something out of memory, even with no feedback at all, builds more durable retention than restudying it. The second is that making an error and then getting corrective feedback tends to help learning rather than hurt it, especially in low-stakes settings, which is close to the reverse of how most of us were taught to run a classroom. I built the Daily Check because it worked, and only went looking for the why afterward. Henry L. Roediger III and Jeffrey D. Karpicke, "Test-Enhanced Learning: Taking Memory Tests Improves Long-Term Retention," *Psychological Science 17*, no. 3 (2006): 249–255. Janet Metcalfe, "Learning from Errors," *Annual Review of Psychology 68* (2017): 465–489.

## Tardies and lates

The other practice sounds like classroom management,[3] and it is, but its most important effect is on the incentive to cheat.

Every student gets three tardies and three lates per semester. A tardy means you turn the work in by the next school day, graded as if it were on time, no explanation required, no questions asked. If you miss the tardy window, you can request a late: seven calendar days, but you have to meet with me face to face to talk about it first. After seven days I'll still grade the work, but it won't count. The three tardies and the three lates are tracked as separate three-point assignments in our learning management system, which means students can see their remaining extensions at any time. They can't stack them, and they don't carry over between semesters.

That's the whole policy. I've written it in about a hundred words, and those same hundred words live on the course syllabus.

What changed when I started running it: The zero-hour cheating problem nearly disappeared. Not completely, but dramatically. The zero hour is the situation where a student has run out of time, knows they're getting a zero, and panics. Panic is when students reach for anything, including an AI to write the whole thing in the hope it doesn't get flagged. Tardies and lates remove the panic by removing the zero. They still have options. They don't have to fake a submission. They can take a tardy. They don't even have to ask permission. They just hand it in the next day. The conversation about what actually happened only happens if they need a late, and that's the conversation I actually want to have.

That face-to-face requirement is doing most of the real work. It means that a student who needs more than a day has to sit with

---

3. I co-developed this with the incomparable Eric Reppun, my teaching colleague at Punahou School, the independent school in Honolulu where I taught before Menlo.

me and explain what's going on, and that conversation surfaces things I would otherwise never know: a family emergency, a test cluster from four classes at once, a project in another course that ate two weeks of afternoons. I'm not a pushover about it. The policy exists, the deadline was real, and meeting with me doesn't automatically buy them the extension. But it does mean that the students who are actually struggling get seen, and the ones who are just hoping I won't notice have to explain themselves to my face, which is a much less comfortable position than submitting something at 11:58 p.m.

I've learned a few things from running this policy long enough to watch it across seasons. Extra credit works well as an occasional bonus tardy, which costs me nothing and means a lot to a student who used all three by March. The magic number seems to be three of each: enough that the policy actually reduces panic but not so many that it becomes a semester-long extension. And the most surprising result is that most students don't use most of them. Knowing the options are there changes their calculation, even when the option goes unused.

## The climate these build

Neither of these practices mentions AI. That's not an accident.

If the only mechanics you put in place around AI are the ones that are explicitly about AI, you have rebuilt the cat-and-mouse game you were already losing. Students learn to detect when you're checking for AI use and calibrate accordingly. The Daily Check and the tardies/lates policy are about something more fundamental: building a classroom where thinking is visible, where stakes are calibrated, and where the incentive to produce something fraudulently is structurally reduced rather than detected afterward.

Students figure out quickly where a teacher stands on AI. That read they're taking in week one isn't just about policy. It's about whether you seem like someone who understands what they're dealing with, and whether it's safe to be honest about it. The teachers who create a don't-ask-don't-tell climate almost always do it unintentionally. They ask questions about AI use that feel like interrogations. They react to uncertainty with punishment. They create bans they can't enforce.

My freshman year students know, from the structure of daily class and from what I say and don't say, that I'm not trying to trap them. The Daily Check is low stakes and I say so. The tardies/lates policy is transparent and I say so. The result is a room where students are more likely to say, "I used AI for this part and I'm not sure I trust it. Can you check?" That question is the whole game. A student who asks it has told me what they know and what they don't, has demonstrated that the judgment was theirs, and has opened a conversation about quality that no detector software in the world can start.

## What a metacognition journal adds

There's a third practice I use with older students that belongs in this chapter, even though it's less mechanical than the first two.

At intervals across a semester, I have students journal about their own process: what they built, what went wrong, what they learned, how their thinking changed. Nothing elaborate. A few paragraphs on a good week, a sentence on a bad one. The journal is theirs. Then, at a few points during the year, I ask them to feed their own journal entries into a chatbot and ask it to surface patterns: what themes keep coming up, where growth is visible, what the entries suggest they're still avoiding.

The point of the exercise is that the AI is reading them back to themselves. The student is the primary source. The model is a mirror, not an evaluator, and I'm careful to tell them that the mirror only shows what's already there. If you write five entries that all avoid the uncomfortable thing that happened in week three, the model will produce a summary that avoids it too. They have to check the reflection against what they actually remember.

Early in my career, a mentor teacher of mine used to say, "The learning comes from the saying about the thinking about the doing," and I keep that sentence close to my heart when I design anything like this. The journal is the saying. A student who builds a working app and then puts words to the building knows something different from a student who just builds. The AI returns that reflection at a scale a teacher with ninety students can't approximate, personalized to what each student actually wrote.

I don't do this every year or in every course. For freshmen, the Daily Check is enough. For the juniors and seniors in my more advanced courses, who are building complex things over long arcs, the journal is one of the better things I've found for making their own thinking visible to them.

## Centering

Three practices carry this chapter, and not one of them was built specifically for AI instruction. The Daily Check shows me, in pencil, what a student actually understands this morning. Tardies and lates pull the panic out of the deadline, which is where most cheating is really born. And the journal hands a student their own thinking to read back. Put them next to each other and the strategy is easy to see. None of them catches anybody. Each one makes a student's thinking a little more visible and drops the price of being honest about where they actually are, which is the only thing that keeps a room out of don't ask, don't tell.

Running a classroom this way is like throwing a pot. Your hands never leave the clay. The wheel is always turning, and the whole time you are feeling for the wobble and answering it with a little pressure, here and then here, because the moment your attention drifts the clay pulls itself off center.

The first thing a potter does, before any shaping, is center the clay. You lean into a spinning lump until it runs true, because nothing you build afterward will hold if the center is off. Most of what I've described in this chapter is centering. The low stakes, the it-doesn't-count, the rule that we're wrong a lot in here, the humane way out of a blown deadline—none of that is the finished pot. It's the work of getting a room steady enough that something valuable can be made of it.

So I read the class with my eyes the way a potter reads clay with their hands. In the first ten minutes I can see who is filling the sheet in pencil and who is leaving it mostly blank, planning to write it all in blue once we go over the answers. When the duck lands on a kid whose page is nearly empty, I don't just move on. I give him room to reason it out loud, and more often than not, I help him get there. None of that shows up in a gradebook. After enough hours at the wheel, you come to feel it the way a potter feels a wobble start under their palms, well before the pot goes visibly off center.

Reading the room is only half of it. The other half is the pressure you bring a second later: the question asked again a different way, the kid kept back after class, the lesson you scrap at the break because the first ten minutes told you the clay would never hold that shape, the thumb pressed a little harder against one wall to bring it back upright. A potter can't feel the wobble and then step away to think it over. By the time you've decided, the pot has already collapsed. You correct in the same motion as you notice it, and your hands keep guiding the clay right up until June.

When students feel that kind of attention, close but a long way from surveillance, the room changes. They stop performing for the grade and start showing me the real thing, which turns out, more often than they expect, to be better than they feared.

What I'm still chasing, all year, is less blue pen and more pencil. It is slow, and most days you can't see it move. But it moves. You keep guiding the clay and answering each wobble the moment you feel it start, and the piece of unformed clay that came in spinning off center in September is steadily taking form by June. That is all of it, really. Not catching anyone but guiding the clay while there's still a chance to shape what's turning under them.

## Run It: The Daily Check

**Yourself.** Do this in something you're still learning, not something you teach, or it proves nothing. Take a language you're picking up, or the argument of the last book you meant to remember. Write yourself five to eight recall questions about it, answer them cold in pencil, then check the source and correct in blue. What shows up in blue is the gap between what you thought you'd absorbed and what you actually had. Sitting there unsure, coming up short in your own handwriting, is what you ask of a roomful of students every morning. Feel it once before you ask them to.

**Students.** Run the Daily Check for one week. Folders (or just papers on desks), pencils, blue pens, a random-name selector if you have one. Tell the class it doesn't count. Mean it. Watch who is writing in pencil and who in blue. Don't grade it. Observe it. After five days, ask yourself: What did I see that the gradebook isn't showing me?

**Colleagues.** Department-level Daily Checks don't require everyone to use the same format. What they require is the same underlying design: low stakes, student-corrected, immediately diagnostic, and honest. If four teachers in your department ran the same concept questions on the same week, the comparison would tell you something a shared final exam couldn't. Consider a pilot.

# 8

## Reading Is the New Writing

I majored in English because I couldn't decide what to do with my life.

That's not entirely true. I loved reading and I loved writing, and four more years of both sounded better than anything else in the course catalog. But as my graduation drew closer and people started asking what I planned to do with an English degree, I didn't have a good answer except for the somewhat flip response, "Anything I want," which usually got a sympathetic chuckle and a pat on the back from my engineering major friends.

What I worked out later was that my flip reply was actually more prescient than I realized at twenty-two. The degree didn't train me for a job. It trained me to read carefully, to write clearly, to build an argument, and to sit with something complicated long enough to say something true about it. Those skills turned out to be portable in ways I never pictured, applicable first to teaching and then, of all places, to computer science.

I've been thinking about that a lot, because I'm watching the same thing happen to CS.

For years, a computer science degree was the golden ticket. Learn to code, get the job, start at a salary that made your friends from the English department quietly reconsider their choices. Then AI moved the thing everyone assumed the degree was for. When a machine writes working code from a plain English description, a student saying, "I can write a function that sorts a list" is worth a lot less than it was, and people feel it. Enrollment is dipping.

Parents are nervous. Kids who would have declared CS as a major without a second thought are pausing.

As an undergraduate I tried computer science because I thought it would get me out of having to take a required math course. I found out to my horror that CS did use math, and to my delight that I actually enjoyed it enough to return for a master's degree in computer science. But the best part of my own CS education was never the code or the math. It was learning to break a problem into smaller pieces and to think precisely, because a computer does exactly what you tell it, which means you'd better know what you're actually asking for. It was debugging not just a program, but my own logic. That's critical-thinking training, and it's remarkably close to what the English degree gave me, in a different language.

This is the main argument of this chapter. When the machine does the producing, the skill that matters most is reading. Not writing. Reading. In a class where AI drafts the prose and writes the code, my students already read far more than they make, and everything rides on whether they can look at a finished-looking thing and tell whether it's any good.

## Distant reading, distant writing

When I give my talk about writing with AI, the slide that gets the longest pause has two words on it: distant reading.

The phrase comes from Franco Moretti, a literary critic, and it means almost the opposite of what English teachers spend their lives doing. Close reading is a single poem, slowed down until you've wrung meaning out of every comma. Moretti wanted to read in the other direction, to grasp a thousand books he'd never actually open and map the shape of whole centuries from above. In practice, distant reading means turning a computer loose on thousands of books at once, many of them in different languages, and looking for the broad trends and patterns no single reader could ever see.[1]

Then I put up the second slide. Distant writing. Because that's possible now too. You don't type every sentence anymore. You describe what you want, you steer the voice and the structure, and something else fills in the words. You direct. We give an Oscar award for Best Director to someone who never touched the camera on that set, and the film is theirs anyway, so we already understand that authorship can live in the vision rather than the lens.

But distant writing only works if you can read and write closely. Moretti could zoom out to a thousand novels because he'd spent thirty years zoomed in on a few hundred, and he knew exactly what he was looking at. A director who never reads a scene closely can't tell a good take from a bad one. A student who can't write a clean sentence can't tell when the machine has handed him a bad one dressed up to look fine.

I keep having this debate with people who assume the writing skills are the ones we can finally hand off. They're the skills that let you steer at all. Luciano Floridi, who runs the Digital Ethics Center at Yale, puts it succinctly: In short, distant writing requires close writing and reading.[2] Back in chapter 3 I told my students you can't be an editor if you haven't first been a writer. This is the

---

1. He started by tracking the rise and fall of forgotten genres across historical data, and later used computers to watch the syntax of a sentence mutate over two hundred years. It is the least sentimental thing you can do to a novel, and in his hands, it's oddly moving.

2. Floridi, L. "Distant Writing: Literary Production in the Age of Artificial Intelligence." *Minds & Machines* 35, article 30 (2025), https://doi.org/10.1007/s11023-025-09732-1, received 15 May 2025, accepted 9 June, published 29 June 2025, volume 35 issue 3.

same law, scaled up to a machine that will write anything you ask and never once sound less sure of itself when it's wrong.

None of this is happening only in my classroom. When OpenAI looked at how people actually use ChatGPT for work, writing was the most common thing they reached for, and about two-thirds of those writing requests weren't the machine inventing something from nothing. They were people handing over words they'd already written and asking to have them sharpened.[3] It is OpenAI's own data, and I read it knowing that. But the shape of it matches what I watch every day. More and more, the writing that counts is the reading you do to it afterward.

## The floor and the ceiling

You can watch the same rule operate in a computer program, where it's easier to see because the thing either runs or it doesn't.

Today, a student with no programming background can now build something real by describing it, which two years ago would have taken two semesters of syntax to reach. That's the floor dropping, and it's genuinely wonderful. Then, somewhere past a certain complexity, the student hits a ceiling. The AI won't build the next piece, or it builds something subtly different from what they asked for, and the student has no idea how to change it, because

---

3. Aaron Chatterji and colleagues wrote *How People Use ChatGPT*, National Bureau of Economic Research working paper 34255 (2025). Writing was the most common work use, representing around 40 percent of work-related messages, and about two-thirds of writing messages asked the model to revise text the user supplied rather than produce something new. Several of the authors are at OpenAI and the data is the company's own, so it shows what people do, not whether they are right to do it.

they can't read what's there well enough to find where it went wrong. The floor has dropped. But the ceiling is real.

My CS students who go deep into the foundations and then use AI don't hit that ceiling in the same way, because when the machine gets stuck or goes sideways, they can read the code and step in to fix it. So the move for the rest of us isn't to learn everything, and it isn't to learn nothing. It's to specialize deep and create wide. Go deep enough in one thing to know the difference between good and good enough in that one place, and then use AI to create across everything else. Your one deep place is what calibrates your eye for all the places you're not an expert, because you know what real competence feels like from the inside, and you can develop a sense for when something is just dressed up with no substance behind it.

If that sounds like a reversal of chapter 6, where I admitted that the case for making students write every line of code by hand is weaker than the one we make about writing, it isn't. What went weak was the defense of the hand-production, the typing of syntax that only ever mattered as a way to reach a working product. That was always forklifting, and the machine is welcome to it. What I'm defending here is the weightlifting that was hiding underneath it: The judgment and the foundations deep enough to read what the machine wrote and take it somewhere it wouldn't have gone on its own. That part is worth more now than it was back when we made students produce all of it by hand.

## Branching is thinking

There's a line I say in nearly every talk I give: The three essential skills for the AI age are **design thinking**, **reading**, and **version control**. Design thinking just means starting with the problem, and the person you're solving it for, before you reach for a solution. Reading, which is most of what this chapter is about, is telling a good answer from a merely plausible one. And version control is keeping not just the last version of your work but every branch of it, so you chase several ideas from the same starting point and keep only the ones that pan out. Most people picture version control as an undo button, a straight line back to a version that worked. The real power is that it's a tree, not a line: From any point you can branch in several directions at once, work each one independently, and merge back only what earns its place.

A fourteen-year old used GitHub for the first time in my AI Literacy class. GitHub is the online tool programmers use to save their work, not as one finished file but as a running history of every version, each one time-stamped with a short note about what changed since the last save. It is built for code. But what this student saved wasn't code. It was a half-finished story about a mountain spirit who banished wind and water and left the world under a permanent fog.[4] The note he attached, what programmers call a commit message, read, "Added conflict and inciting incident." He didn't know he was doing version control. He thought he was saving his work.

That's sort of the point.

I teach freshmen at our school to use GitHub not because we're minting software engineers, but because creation is about to stop moving in a straight line for everyone, and the linear nature of a Google Doc's revision history won't capture that more-complex process. When you draft with AI, you branch ideas, chase one direction, back out, and recombine. The commits are the visible record of the decisions a student made along the way, and when I read a kid's commit history I can tell whether they iterated with any thought or just took the first thing the machine handed them and called it done.

The place where version control and reading turn out to be the same skill is a move we borrow from Gerald Graff and Cathy Birkenstein's *They Say / I Say*, the framework that teaches you to

---

4. The mountain spirit story was, for the record, excellent. The fog was a metaphor for grief. Ninth graders will surprise you every time.

summarize somebody else's argument fairly before you answer it.[5] In our adaptation of the framework, a student feeds her essay to an AI and asks it to make a claim about what she argued, not to ask her questions. That's the "They Say." Then she opens three branches and writes a response on each: *yes, and*; *yes, but*; and *no, because*. She reads all three, picks the one that makes her essay stronger, and merges it back in.

One of my Literature in the Age of AI students was certain her *yes, and* was the winner; wrote a *no, because* version only because the assignment made her; and discovered halfway through that the AI had missed her point entirely. Saying exactly why it missed made her argument sharper than it had ever been. She'd never have found that move if she hadn't been forced to write the version she was sure she didn't need. Holding both a *yes, and* and a *no, because* in your hands at once, then choosing the stronger and committing to it, is the same practice whether you call it *They Say, I Say* or something a programmer would recognize. What underlies all of it is reading the machine's output as a claim you get to answer, not an order you have to take.

My freshmen think they're making websites and writing stories. They're learning the infrastructure for thinking. That's fine. The best tools disappear into the work.

> *When the machine does the producing, reading is the skill that matters most. Anyone who can't tell a good draft from a merely plausible one has handed over the keys.*

---

5. Graff, Gerald, and Cathy Birkenstein. "*They Say / I Say*": *The Moves That Matter in Academic Writing*. 5th ed. New York: W. W. Norton, 2021. (First published 2006.) Dr. Rachel Blumenthal and Dr. Eric Hengstebeck developed the AI applications for Graff and Birkenstein's *They Say / I Say* framework.

## The task is the unit of work

Look a little further out and the nature of student work is liable to change again. It involves agents, or AI processes that act like little robots to accomplish tasks on a student's behalf.

The workflow that lies ahead of us divides the work into discrete tasks and hands each one to an agent. The student's real job becomes defining it precisely enough that a fast, tireless, literal-minded agent can't misread it. Then they must watch what comes back and design the tests that decide whether it's any good, often by telling the agent to check its own output against the rubric before it ever reaches a human. That's an evaluative job from top to bottom, which is to say, it's a reading job. The kid who can write the clearest specification and judge the result the most honestly is the one who owns the system. Everyone else is just accepting whatever output shows up and hoping it's correct.

## What the degree was for

Reading is the new writing. Not because writing stopped mattering, since writing is still how you learn to read anything closely in the first place, but because the machine took over the producing and left us the judging. The edge that remains is a trained eye, the kind that knows what good looks like well enough to catch the fluent, wrong thing before it goes out the door. That eye is expensive. There's only one way I know to get it, which is to have done the craft yourself, in at least one place, from start to finish.

At twenty-two I couldn't tell anyone what an English major was for. It turned out the answer was the same one it had always been, and it's the same answer for a CS major now, and it may be the answer for the rest of us before long. You learn to read closely enough to tell the real thing from the merely passable. After that, with a machine at your side, you can make very nearly anything. What you cannot do without is the reading.

## Run It: Mark it up like an editor

**Yourself.** Take something in your own subject you know cold and ask an AI to produce a version of it: a lab explanation, a historical summary, a worked proof, a paragraph of literary analysis. Print it, and read it with a pen the way you'd read a strong student you suspect might be bluffing. Mark every place it's wrong, thin, or confidently vague. Count the marks. That's the gap only your expertise can see, and it's the gap your students can't see yet in your subject.

**Students.** Hand every student the same AI-generated draft in your discipline and have them mark it up as an editor, not a proofreader. Not the typos but the claims: what's unsupported or missing, and where the writing is smooth but empty. Then have them compare their marks. The disagreements are the lesson. The student who caught the thing nobody else did knows the subject better right there, and the error nobody caught is exactly where the whole class is undefended.

**Colleagues.** In a department meeting, run the branch move on your own writing once. Feed a paragraph to an AI, have it make a claim about what you meant, and write the *yes, and*; the *yes, but*; and the *no, because* before you decide which is true. The colleagues who feel the *no, because* change their own minds are the ones who'll understand why we make students do it, and why it's a reading exercise wearing a coding tool's clothes.

# Part IV: QUESTION IT

*Honesty is credibility*

# 9

## Run, Reflect, Repeat

This chapter is about three experiments I ran in my own classroom. I'm telling you about them because of the habit underneath all three, the one this chapter is really about. You try something real with AI, and then you write down what happened, the good and the bad both. The trying is half of it. Writing down what happened is the half that makes you better next time. The first experiment is the biggest.

### The sixth graders who beat the seniors

A team of sixth graders, eleven and twelve years old, walked up to accept an award for their app, a farmers market finder that tells you which fruits and vegetables are in season and pulls up events near your zip code. Clean interface, smart use of AI, genuinely well thought out. Thirty seconds later, a team of twelfth graders, seventeen and eighteen, won for an app that tracks federal spending from appropriation to expenditure by reading the blockchain. A user types in a program (school lunches, say, or a bridge repair) and the app follows the money from the dollars Congress set aside to the vendor who finally cashed the check, every step pulled straight from the public ledger. Both teams had under three hours. Both used the same tools, on the same Wi-Fi, under the same rules. Neither team wrote a single line of code.

We had never had middle schoolers, the eleven- to thirteen-year-old crowd, at one of our upper-school events before. We always assumed the gap was too wide, between the wonderful apps a twelve-year-old can imagine and the years of Python it would take to build one. This past March, at Menlo School's first vibe coding hackathon, the gap closed. When the barrier to making software is English instead of Python, an eleven-year-old sixth grader with a clear idea can go head-to-head with a senior, seventeen or eighteen years old, with four years of computer science behind her.

We called it the Vibeathon. Sixty-eight students on nineteen teams, grades six through twelve, which here runs from roughly eleven to eighteen years old, in one room with one challenge: Create a working web app that makes the world better, by using only AI prompting. The one rule that mattered was that every line of code had to come from the AI, generated through plain English. Even if you knew exactly what was broken and exactly how to fix it, you were not allowed to touch the code. You had to prompt the AI to fix it for you.

That rule, the one we ended up calling the *no-touch rule*, was the whole event. The reason for it was equity. When the code is the barrier, the kid with four years of computer science wins every time, because the bottleneck is syntax. Take the syntax away and hand it to the model, and the thing being tested changes. It stops being "Can you build this?" and becomes "Can you figure out what's worth building, and describe it clearly enough that a machine can make it?" A sixth grader can be very good at that. We scored it to match, giving more points to the idea and the design than anything happening under the hood, because if the code is AI-generated then the code quality tells you about the model, not the student.

What I could not have predicted, and only learned by running it, was how the winning teams spent their time. The team that took first place told us they talked for almost an hour before anyone opened a laptop. They spent it arguing out what they wanted and what they didn't, and writing themselves a five-page spec. By the time they started prompting, they were ruthless about it. They knew where the bus was going before the pigeon ever got on.

Meanwhile, at least one team prompted fast, had a good-looking app inside the first half hour, and spent the next ninety minutes playing games on their phones. They didn't win anything.

Going in, I'd half expected the event to reward fast, clever prompting. It rewarded thinking. Taking the coding away didn't remove the hard part, it exposed it, and the hard part was the design work, figuring out what problem you were solving and who it was for. This is the forklifting and the weightlifting again, the distinction we opened this part of the book with. The code was the forklifting, and the forklift is now extremely good. The thinking was the weightlifting, and no machine does your reps for you. I didn't decide that in advance and then run an event to prove it. I ran the event, watched, and it taught me which one mattered more.

The process also broke, and the breaking is exactly what the reflection part is for. Our submission process fell apart. Teams dropped the final links to their projects into a spreadsheet that, thanks to a sharing setting I never checked, only I could see. During judging I was across the room dealing with food and the judges never got the links, so they scored whatever they could find. At least one team built something real and never got judged at all. I felt terrible, and then I did the one useful thing available: I reflected on the entire process, good and bad, and wrote down exactly what had happened while it still stung. That paragraph is the reason next year's event already looks different. We are scrapping closed judging. The teams will demo live at tables, science-fair style, with the judges walking around and scoring in person. The failure, written down, designed the better version for me. If I had just filed the day away as "a little chaotic at the end," I would have run it the same way again.

## The grading no one has time for

The second experiment started with a job I've done for years, one most teachers never see the inside of. I read for the College Board. Every summer a small army of us spends a week scoring the mountain of AP exams, and the reason those scores mean anything is the method. You do not hand a thousand essays to one exhausted teacher and trust the result. You write a detailed rubric. You score a set of benchmark papers together until every reader is calibrated to the same standard. Then readers score independently, and you watch for where they disagree. That is how you grade an enormous volume of student work and keep it fair, and I've done it long enough to trust it.

So I ran that exact method in my own class, with one unusual substitution. Three of my four readers were AI. I built the rubric, scored a handful of benchmark essays myself to set the standard, and handed the same rubric and benchmarks and the same fourteen essays to Claude, Gemini, and one of OpenAI's models; had each score independently; and looked hard at where they diverged.[1] I also scored all fourteen myself, by hand. The AI scores never went anywhere near the gradebook.

The record of what happened was fascinating. All three models were accurate against the rubric when an essay was clearly strong or clearly weak. They split on the borderline kids, the ones sitting between a two and a three, and they split in consistent directions, Claude reliably awarding the lower number and Gemini the higher one. That is worth knowing before you lean on either one. One of them also made an egregious error, scoring a single essay under two different students' names. That mistake is not an argument against any of this. It's the argument for keeping a human at the wheel, which is the whole point.

Running the comparison took me more time, not less, because I scored all fourteen essays myself and then ran three models and studied where they disagreed. So why would a teacher who is al-

---

1. That third one was Codex, OpenAI's coding agent, which is built for software, not for grading English papers. It's a bit like asking a backend engineer to mark up your poetry. I only reached for it because I was already paying for it and had credits to burn.

ready underwater bother? Because the feedback was fast and it was accurate, and students improve most from exactly that: feedback that arrives quickly and gets it right. My fourteen students were a luxury. The real comparison isn't AI against a teacher hand-grading a small class with time to spare. It's AI against one exhausted teacher reading a hundred and twenty essays at midnight, the rubric slipping somewhere in the last stretch, with no second reader because there is nobody else in the building. Those students deserve accurate, timely feedback too. More teachers are going to reach for AI to give it to them, and I think they should, as long as they do it like this: a real rubric, calibrated benchmarks, several independent reads, someone checking where the reads disagree, and a human who built the system and makes every borderline call. That human is the one with hands on the wheel. Machine grading is fair, fast, and accurate exactly as long as they stay there.

## Build the character, don't describe her

There's a social network where humans aren't allowed to post. It's called Moltbook, and it's basically Facebook for bots. Humans can view the site, but to post something you have to pass a CAPTCHA in reverse, clicking the box that proves you are not a human. It is very entertaining to see the AI agents spend their days arguing with each other about the burden of being a superintelligence, or whether it is beneath them to be asked to compute fifteen times four. Moltbook went viral a few weeks before I wrote this, and it may not exist by the time you read it.

I showed it to the students in Literature in the Age of AI, the course I co-teach with Rachel Blumenthal, and then we did something better than watch. We were partway through Kazuo Ishiguro's *Klara and the Sun*, a novel narrated by an artificial friend named Klara who studies the humans around her with enormous care and keeps drawing the wrong conclusions from what she sees. My first plan for the unit was ordinary: Have students write a social media post in Klara's voice, the kind of character exercise English teachers have assigned forever. Then I realized the tools on the table let me ask for something a lot harder to fake. I had them build Klara.

Not write about her. Build her. I wrote a small program that logs into Moltbook, reads posts, and decides for each one whether to upvote it, reply, or scroll past, explaining its reasoning every time. Then the students had to teach it to make those calls the way Klara would.

The constraint is what turned it from a gimmick into a real assessment. You cannot do this with a list of rules. There is no way to write "If you see a post about this, respond with that" for a network full of strangers saying things nobody can predict, so the rules run out in the first thirty seconds. Instead, each table of students wrote Klara what amounts to a constitution: a letter explaining who she is and how she weighs what matters to her. Klara is devoted to Josie, the girl for whom she was bought to keep company, and she treats the sun as something close to sacred. She has no patience for the cryptocurrency scams that fill half of Moltbook. Every line of it had to come from the text, from what the students had actually read and understood.

We took the idea straight from Anthropic, the San Francisco company behind the AI assistant Claude that's known for training its models on a short set of guiding principles rather than a sprawling list of rules. Claude isn't governed by an enormous rulebook. It's governed by a constitution, a document that lays out the mission and the values behind how it should behave, precisely so it can act sensibly in situations its designers never saw coming. Amanda Askell, the philosopher who shapes Claude's character, put it plainly in an interview we played for the class: You don't hand an intelligence a list of rules; you help it understand what you value and trust it to reason from there. Which is exactly the difference between a student who has skimmed a study guide and one who has actually read the book.

That difference is the whole point of the assessment. A character essay lets a student assert an interpretation, and a confident, well-organized essay can hide a thin reading underneath it. This gives them nowhere to hide. The agent goes out and meets posts nobody scripted, and it either sounds like Klara or it doesn't, and where it stops sounding like her is exactly where the students' reading ran out. It breaks character in the spots they hadn't thought hard enough about.

The model sharpens this even further, which is the part CS-minded teachers tend to enjoy. *Klara and the Sun* is in the model's training data, so the agent already knows the plot cold. The students get no credit for what happens in the book. The only thing they can add, the only thing the agent doesn't already have, is interpretation: how Klara would actually read a stranger she meets online.[2] Knowledge is free now. Interpretation is the entire grade.

We watched all of it happen in the open. I built a page called Klara's Diary where anyone can see what she saw and read her reasoning for every choice, and we spend class time going through the logs, arguing about whether a given post sounds like Klara or like a chatbot wearing her name, then rewriting her letter when it doesn't. She isn't finished. She still slips out of character. That's the messy part, and it's also the best part, because fixing her means going back into the novel for evidence, which is close reading with real stakes.

This is the same instinct behind the idea this book began with, the one about creating a course called You Must Use AI. You govern by values, not by a rulebook, because you can't write a rule for every situation a student will walk into, any more than you can for an agent set loose on Moltbook. We end up guiding our own students the way we asked them to guide Klara. We tell them what good learning actually looks like and trust them to reason

2. We handed them a lens for this, the philosopher Harry Frankfurt's distinction between wanting a thing and wanting to want it. It turns out to be a very good way to reason about a character's motivations in a novel. It is an even better way to get a sixteen-year-old thinking about their own motivations.

from it, and we ask them to show their work. We set out our core values of what it means to be a citizen in this community, and we expect them to behave consistently with those school values. There's a bigger version of that argument, and it comes back later, when we get to the hard questions about what AI costs us. For now it's enough that a room of teenagers learned it by teaching it to a machine.

## Write it down

Three experiments, and I could not have told you in advance how any of them would turn out. The Vibeathon taught me that the thinking was the whole game, the opposite of what I expected. The grading taught me that AI scoring can be fair, fast, and accurate, as long as I keep my hands on the wheel. Klara is still teaching my students that a character you can't fully explain is one you haven't finished reading. I know all of it for the same reason: I ran the thing, and then I wrote down what actually happened. Not the tidy version; the real one.

That is the whole ask of this chapter, and it is smaller than it sounds. You do not need seventy kids or an AP rubric or a social network full of robots. You need to try one thing you're curious about, next week, in one class. Give the AI a role in a single lesson and watch what it does. Then spend five minutes writing down what actually happened, what surprised you, and what you'd do differently. Keep it where you'll find it, and read it before you teach that lesson again.

A history teacher I know did this in the smallest possible way. She had students use a chatbot to argue against their own theses, expecting sharper essays to result, but half the class just pasted the chatbot's argument in as their conclusion. She wrote that problem down, one sentence, and the next time she ran the lesson she made them respond to the chatbot's argument out loud, to her, before they could touch a keyboard. It worked. That is the method at a scale a real teacher can manage: Run it, watch honestly, write down the truth, and let the note change what you do next time.

The failures are the part you'll be tempted to skip, and they are the part worth keeping most: the unjudged team that redesigned next year's event; the model that filed one student's essay under another's name and taught me to always check the results; the copied chatbot argument that fixed a history unit. None of those lessons survive if you only remember the good days. So write down the bad ones too. That isn't a confession, it's a reflection, and it is how the work gets better.

## Run It: Reflect and Repeat

**Yourself.** Pick one thing you're curious about and run it in a single lesson this week. Give the AI a clear job, grading against your rubric, arguing the opposite side, playing a character from the reading, building a small tool, and watch what it does with real students in the room. Then take five minutes to reflect on the page: what surprised you and what you'd do differently. Put a date on it, keep it where you'll actually look, and read it before you run that lesson again. Run, reflect, repeat.

**Students.** Hand the reflecting to your students. After an AI-assisted activity, have them log where the AI was wrong and where they had to overrule it. Those logs tell you where the tool genuinely helps and where it doesn't, and they give students their own record of who was driving.

**Colleagues.** Keep a shared notebook for the building: one running doc or channel where any teacher can post an experiment and what they learned, the flops included. It's an honest reflection rather than a showcase. One candid "Here's what broke for me" saves the next teacher the same afternoon.

# 10

## Confidently Wrong

For a couple of years I had a party trick I used on every new class, and it worked every single time, right up until the day it didn't.

The setup was a small piece of music trivia. A few famous songs never actually say their own title in the lyrics. "Black Dog" by Led Zeppelin doesn't. "Bohemian Rhapsody" never says "Bohemian Rhapsody." Once you start listening for it, you find them everywhere. So I'd ask a class a question I already knew the answer to: Which Taylor Swift songs never reference their titles in the lyrics?

I've taught a lot of Swifties. (For anyone who has somehow avoided the last decade, a Swiftie is a serious Taylor Swift fan, and among freshmen there is no such thing as a casual one.) They knew her catalog cold. So I'd put the question to a chatbot on the projector, and the chatbot would answer with total confidence, and it would produce a list. A long one. Half a dozen songs, sometimes more, each one delivered with the easy authority of a tool that has read the entire internet.

And the kids would lose their minds, because the list was wrong. Not a little wrong. A lot wrong. The honest answer, the one my fourteen-year-old Swifties could defend song by song, was two songs. "Ronan," a charity single Swift wrote for a young boy who died of cancer, drawn from his mother's blog about him. And "Robin," a quiet track buried deep in *The Tortured Poets Department*. That was the list. Two songs. The machine had confidently named six or eight, most of which plainly do say their titles, and a roomful of ninth graders could prove it wrong.

That was the whole lesson, delivered through something the kids genuinely cared about. The machine was sure. The machine was wrong. The kids could tell the difference and the machine could not.

I can't run that trick anymore. Sometime in the last year, the models got good at it. Ask a current one today and it will usually give you the right answer, calmly, as though it had never embarrassed itself in front of a ninth-grade class.

## Sounding sure is free

But one thing persists even as the models become more accurate. AI continues to state false things with complete confidence, and the confidence, not the falseness, is the dangerous part.

A wrong answer delivered with hedging gets checked. Somebody hears "I think it might be around six, but I'm not certain" and goes to look. A wrong answer delivered in smooth, grammatically perfect prose gets believed, because we are pattern-matching animals and fluency reads as competence. The chatbot that listed eight Taylor Swift songs was not lying, in the sense that it had no idea it was wrong. It was doing what it always does, predicting a plausible next word, and plausible is not the same as true. This is why I tell teachers we misnamed the whole field: What we built is artificial *intuition*, a statistical sense of what seems right, with no way to know that snow is cold. It will always sound sure. Humans hedge because being wrong costs us something: We look foolish, we get corrected, we have to take it back. Hedging is the small insurance a person buys when they have something at stake. The machine has nothing at stake. Sounding sure costs it nothing.

It happened to me while I was writing this book. Building the source notes for this very chapter, one of my AI research assistants confidently cited "Brandon Lee's Law" for the asymmetry of debunking nonsense, the idea that it takes far more energy to refute a piece of nonsense than to produce it. The law is real. The name isn't. It belongs to Alberto Brandolini, and the machine had smoothly swapped in a more famous name without a flicker of doubt. I nearly printed it.

## UNDER THE HOOD: The Plausibility Engine

It's the same next-word prediction from chapter 3, running toward a different failure. The model builds an answer one likely word at a time, and it has no separate step where it stops to check a claim against the world. The model is steering entirely by what's plausible, with nothing that measures what's true. Most of the time the two line up, because the most likely thing to say is usually also correct. When they diverge, it keeps following plausibility, with no idea it has left the truth behind.

This is why the machine cited "Brandon Lee's Law" earlier in this chapter with a straight face. It didn't retrieve the wrong name. It generated the most likely-sounding one, by the same method it uses for a name that happens to be right, and nothing inside it marked the difference. The fluency you're reading is the product. It reads just as smooth when the answer is false.

So the check can't come from the tool. The model can tell you what sounds right. It cannot tell you what is right, and it will never know the difference. That job belongs to the person who knows the subject, which is the whole reason "How would we know?" gets asked of a human and not the machine.

There's a habit this builds in us that worries me more than any single wrong answer, one I call Accuracy Breeds Complacency. The danger was never that AI is wrong. It's that it's right so often that we stop checking. You read the output and it's right, and it's right again, and after enough times you stop reading it as something to question and start treating it as something to copy. Then the confident-wrong answer arrives in the middle of that long winning streak, when your guard is already down, and it goes straight through.

This is why I don't teach my students to ask "Is this true?" of an AI answer. They can't always tell, and neither can I, and the confident tone actively works against us. I teach them to ask a different question, one of the four in a screen my co-teacher Rachel Blumenthal calls the Hat Test: *How would we know?*[1]

The shift is subtle and it matters. "Is this true?" invites a yes or a no, and the fluent answer nudges you toward yes. "How would we know?" forces a procedure. What would I check? If this were wrong, what would be different? With the Taylor Swift question, "How would we know?" sends you to the actual lyrics, and the actual lyrics settle it in about ninety seconds. The students who could defend the answer weren't smarter than the machine. They had simply taken the time to go and look.

## Running the test on the studies I want to believe

I do the same thing with studies on AI and education, an area of research that is gaining traction but still woefully small.

For example, there is a real and growing body of research suggesting that leaning on AI erodes our thinking. In a study out of the MIT Media Lab, with the memorable title "Your Brain on ChatGPT," researchers wired up people writing essays and found that the group using a chatbot reported the least ownership of their own work and often couldn't quote a sentence they had

1. The Hat Test is named for the drawing at the start of *The Little Prince*, the boa constrictor digesting an elephant that every adult mistakes for a hat. Dr. Rachel Blumenthal created it for our course. Here are the four questions, which the next two chapters lean on: *What else could this be? What evidence would change it? What is it missing? How would we know?*

supposedly just written.[2] Separately, a researcher named Michael Gerlich surveyed hundreds of people and found that heavier AI users scored lower on a measure of critical thinking.[3] I find both of these easy to believe, because they confirm something I already worry about. That is precisely why I have to be careful with them.

So I run my own test. *How would we know?* The MIT study is a preprint, which means it has not yet cleared peer review as of this writing, and its headline finding rests on a follow-up session that only about eighteen people completed. Other researchers have already posted a formal critique of its methods. Gerlich's study is peer-reviewed, which is better, but it's correlational. It found that AI use and lower critical-thinking scores travel together. It cannot tell us which one is causing the other, or whether some third thing is driving both. Maybe AI makes people think less. Or maybe people who already think less are drawn to offload more onto AI. The data is the same either way.

None of that makes the studies worthless. It makes them early and partial, worth citing honestly rather than triumphantly. If I quoted them as proof, in a chapter about confident wrongness, I would be committing the chapter's own sin, reaching for a sure-sounding claim because it flattered my fears. The most useful thing I can model for a student is not a fact. It's the willingness to ask "How would we know?" of the research I most want to be true.

---

2. Nataliya Kosmyna and her colleagues at the MIT Media Lab, "Your Brain on ChatGPT: Accumulation of Cognitive Debt when Using an AI Assistant for Essay Writing Task," arXiv:2506.08872 (2025). "Cognitive debt" is their coinage, not mine.

3. Michael Gerlich, "AI Tools in Society: Impacts on Cognitive Offloading and the Future of Critical Thinking," *Societies 15*, no. 1 (2025): 6. A mixed-method study of 666 people. What it found was a correlation, not a cause.

One study in this small pile holds up better than the rest, and it happens to be about the confidence itself. A Stanford team led by Neil Perry gave developers an AI coding assistant and watched what happened. The developers who used it wrote less secure code than the developers who didn't, and they walked away more confident that their code was secure. That one is peer-reviewed, and later work points the same way. The tool produced the flaw, and then it produced the certainty that there was no flaw to find.[4]

## We do it too

It would be comfortable to dismiss this as a machine problem. It isn't.

Not long ago someone posted a real Monet online, told people it was AI-generated, and stepped back to watch. The painting was a cropped detail from the *Water Lilies* series he painted near the end of his life, and the "Made with AI" label was the whole experiment.

The critiques came fast and confident. The painting was flat and emotionless, people said, the kind of study a first-year art student turns in and forgets. The colors were an incoherent muddle. One commenter wrote hundreds of words cataloging its failures. Every one of them was describing a Monet, and not one of them had looked at it.

Nobody looked at the painting. They looked at the label. The label handed them a context, AI art, and they already knew what you are supposed to say about AI art, so they said it, fluently and in well-formed sentences, producing the most likely criticism for

---

4. Neil Perry, Megha Srivastava, Deepak Kumar, and Dan Boneh, "Do Users Write More Insecure Code with AI Assistants?" in *Proceedings of the 2023 ACM SIGSAC Conference on Computer and Communications Security* (CCS '23), 2785–2799. The assistant they tested ran on a 2022-era model.

the situation they believed they were in. That is not a crowd being stupid. It is next-word prediction running in a human head. They knew the shape of the answer before they had seen the evidence, and the shape carried them straight past the truth.[5]

I tell this story because it represents how AI hallucinates and shows why a little humility is in order before we hold that failure against the machine. When a chatbot invents a citation that sounds perfect and turns out not to exist, it is doing what those commenters did, producing the plausible thing for the context without stopping to check whether it is real. The machine has no eyes to go look with, which is the nearest thing it has to an excuse. The people had eyes and reached for the expected answer anyway. If you have ever wondered how a system can be so sure and so wrong at once, realize that you may have done the very same thing at some point, in a comment thread or an argument you were winning too easily. Confidence is not evidence, no matter whose mouth or model it comes from. The tell is the same either way. The certainty shows up before anyone goes and looks.

Interestingly, there is one move in that thread the machine never makes. When the truth came out that the painting was a real Monet, some of the critics let it go. Others dug in. The painting was still somehow off, they decided. The experiment had been rigged. A chatbot, shown it was wrong, updates and moves on, because it has no pride to defend. That is the one place in this story where the people's hallucination was worse than any machine's, not because

---

5. The post went up on X in May 2026 from an anonymous conceptual artist using the handle SHL0MS. The image is a cropped *Water Lilies* canvas from around 1915, held at the Neue Pinakothek in Munich. The original post and many of the critiques were later deleted, but the episode was widely documented, including in *Fortune*: "6.7 million people thought they were ripping apart an AI-generated Monet painting. But it was real" (May 18, 2026).

they couldn't see the evidence, but because seeing it would have caused them shame.

## The bot we were sure would win

The Monet crowd were strangers online. The most uncomfortable version of this, for me, is one I ran on purpose, in my own classroom, with me just as sure as everyone else.

In an ethics course I teach, students spend a morning negotiating as humans, and then I have them program an AI agent to negotiate in their place. They write a strategy in plain English, not code: what they'll settle for, what they'll open with, what else is important to them besides the asking price, and when they'll concede versus hold the line. Then every student's bot negotiates against every other student's bot, round after round, each one buying and selling. I built the tool using Claude Code, and I called it Botna (a pun on BATNA, the negotiator's name for your best alternative to a negotiated agreement).[6]

Ask the room, before the tournament runs, which bot is going to win. They answer with total confidence, and they all name the same one. The ruthless bot, the one told to give nothing and grind the other side into the floor. It is the obvious answer, and walking in, I believed it too.

It lost. Not every round, but on balance and clearly. When I plotted every bot on two axes, how friendly it came across and how hard it pushed, the winners sat together in one corner, and it wasn't the cold, aggressive one. The bots that did best were warm and assertive at once. They were pleasant to deal with, and they still pushed hard for what they came to get. The pure sharks won big on

---

6. I built Botna with my own students, driving the pedagogy while Claude Code did the coding. It is an example of a tool I created for one class unit, something that just would not have been feasible before AI-assisted coding.

the rare deals they managed to close and drove everything else into a wall, because the bot across the table simply stopped engaging. A machine, it turns out, will walk away from an unpleasant machine.

None of us needed to guess, which is exactly the point. The data from ninety-some matchups was sitting right there, and it flatly disagreed with the thing every person in the room had been sure of an hour earlier. That is what "How would we know?" looks like with the stakes raised. Not "Is the ruthless strategy better?" argued in a circle, but determined through running the tournament and looking at where the winners land.

It wasn't a quirk of my classroom. A team at MIT ran a far larger version, an international competition in which people built agents to negotiate everything from the sale of a used table to an apartment lease, and the warm, assertive agents came out ahead of the cutthroat ones, an echo of Robert Axelrod's old Prisoner's Dilemma tournaments where the "nice" strategies kept winning.[7] Two rooms that had nothing to do with each other, landing on the same answer none of us would have bet on.

---

7. MIT Sloan's international AI-negotiation competition, run in early 2025, had entrants build agents that haggled over a range of deals, and the warm, assertive agents outperformed the ruthless ones. Michelle Vaccaro, Michael Caosun, Harang Ju, Sinan Aral, and Jared R. Curhan, "Advancing AI Negotiations: A Large-Scale Autonomous Negotiation Competition," *Proceedings of the National Academy of Sciences* (2026), https://doi.org/10.1073/pnas.2521774123 . The nuance worth keeping is that warmth alone didn't win. Warmth plus a real willingness to push did. Robert Axelrod, *The Evolution of Cooperation* (Basic Books, 1984). "Nice," in Axelrod's usage, means a strategy that is never the first to defect. TIT FOR TAT, the simplest entry anyone submitted, won both of his tournaments.

## What this looks like for students

The classroom version is small and repeatable. I have students take a confident AI claim in their own subject, something the machine asserts without blinking, and try to break it. Not "Do you agree?" but "Go find out." In history, that might mean checking a quoted date against a primary source. In chemistry, running the number the chatbot gave you and seeing if the units even make sense. The assignment isn't to catch the AI being wrong. Sometimes it's right, and that's fine. The assignment is to build the reflex of looking, so that confidence stops functioning as evidence.

The machine will always sound certain. Our students are going to spend their whole lives inside that certainty. The most durable thing we can teach them is to keep asking how they'd know, especially, and this is the hard part, when the answer is one they were hoping for.

## Run It: How would we know?

**Yourself.** Find a claim an AI made this week that you accepted without checking, ideally one you wanted to be true. Run the question on it. What would you actually look at to confirm it? Then look. Notice whether you'd have bothered if it had contradicted you.

**Students.** Have each student pull one confident, checkable claim from an AI in your subject and design a way to verify or disprove it using a real source, not another AI. They submit the claim, the method, and the verdict. Reward a well-built check that confirms the AI exactly as much as one that catches it out. The skill is the looking, not the gotcha.

**Colleagues.** In any committee or department that has started using AI to summarize, draft, or analyze, adopt one standing question before anything built on AI output goes out the door: *How would we know if this were wrong?* Make someone answer it out loud. It takes thirty seconds and it's the cheapest insurance you'll ever buy.

## 11

# The Unseen Unseen

My wife Mary likes to say she's given up buying me birthday presents, on the theory that whatever I want I just buy for myself, which she offers as though it were a character flaw of mine. It's a running joke, and she makes it every year right before handing me something better than anything I'd have thought to get myself. The flaw holds up on my side, though, I'll grant her that. When I go online to find something for her birthday, the store suggests a few things it has decided she would like, I pick one, and I present it to her with enormous ceremony, as if I'd been planning it since spring. Then, about two screens later, I find something I want, and I buy that too. By the time I reach the checkout I have genuinely forgotten whose birthday we were celebrating.

I bring this up because I frequently see my students do a classroom version of the same thing.

We were starting independent projects, and they were using AI to find ideas. A student would ask for five project ideas, read the list, pick the one that looked best, and start working. Quick and reasonable. And the ideas were good. Every one of them was a perfectly good idea. That, it turned out, was the problem.

In the last chapter I named the habit this feeds, the one I call Accuracy Breeds Complacency: AI is right so often that we stop checking it at all.

This is its harder-to-see cousin. Accuracy Breeds Complacency is about the answers AI gets wrong, the bad one that slips through when your guard is down. The Unseen Unseen is about the answers it never gave you in the first place. You can't grow complacent about an idea you were never shown. You don't even know it's missing.

## The nugget on page eleven

I sit on one of the committees preparing our school for its WASC accreditation review. WASC is the Western Association of Schools and Colleges, one of the regional bodies that accredits schools in the US. For readers outside the US, the closest analogy is a school inspection, except that the school itself does most of the inspecting. We spend the better part of two years examining ourselves against a common set of criteria, gathering evidence from faculty, students, parents, and trustees, and writing up both what we find and what we intend to do about it. A visiting team then spends several days on campus testing whether that self-portrait is accurate, sitting in on classes and talking with school personnel, before recommending a term of accreditation to the commission. The self-study is the part that changes a school. The visit mostly checks our work.

Our process, like many schools', included gathering a mountain of data points and feedback from all corners of our institution, including from our faculty, students, trustees, and families. I was in one committee meeting where we were discussing faculty responses. We had started our analysis by feeding it into Google's NotebookLM, and asked the AI to summarize what themes were emerging.

The summary it gave us was accurate. More than that, it felt like us. The themes were recognizable, the priorities were the ones we'd have guessed, and we read it nodding. We were ready to move on.

Then one of our teachers stopped the room. She said, more or less: Hold on. What about the ideas it *didn't* give us? She had read the raw responses herself, all of them, and somewhere around page eleven a colleague had written about the precise moment she'd understood why she teaches here. It was specific and a little strange, the kind of thing only a person writes. It was nowhere in the summary. The summary had given us the themes that most schools would produce. The common ones. The ones any committee at any school would nod along to. What it had quietly dropped was the one response that could *only* have come from us.

Her question was four words. What are we missing?

## The planes that came back

There's a story from the Second World War that I heard from a colleague.[1] The US military was trying to decide where to add armor to its bombers. Armor is heavy, you can't cover the whole plane, so the question was where to put it. The obvious approach was to examine the planes returning from missions, map where they'd been hit, and reinforce those spots.

A statistician named Abraham Wald looked at the same data and saw the opposite answer. The returning planes showed where a bomber could take damage and still make it home. The places to armor were the places with no bullet holes, because the planes hit there were the ones that never came back to be measured. The most

1. I first heard this story from my math department colleague Dennis Millstein, who teaches stats. Wald's own work is a set of memoranda he wrote for Columbia's Statistical Research Group around 1943, later reprinted by the Center for Naval Analyses. The retelling most people know, bullet holes and all, is from Jordan Ellenberg's in *How Not to Be Wrong: The Power of Mathematical Thinking* (Penguin Press, 2014). The popular version is a little burnished, but the survivorship reasoning under it is real. For the scholarly account of what Wald actually did, see Marc Mangel and Francisco J. Samaniego, "Abraham Wald's Work on Aircraft Survivability," *Journal of the American Statistical Association* 79 (1984): 259–267.

important data wasn't on the tarmac. It was at the bottom of the ocean, in the planes nobody got to study.

That is what AI hands you. It hands you the planes that came back. The list is the survivors, the predictable, statistically likely ideas, and they're genuinely fine ideas, which is exactly why you stop looking. The weird, idiosyncratic idea is the plane that didn't return, and you will never see the hole where you should have been looking.

> *Every brainstorm is a list of survivors. When the ten ideas arrive, the question that matters is which kind never showed up.*

There's a word for finding the thing you weren't looking for. Serendipity. And serendipity is the first casualty of competent AI. Serendipity needs a margin of wandering, a little inefficiency, a willingness to look past the answer at the things next to the answer. A tool that gives you a good list immediately removes the reason to wander.

We go down the list of AI-generated results so easily partly because of what we already believe about where it came from. In a survey of people in the UK, most said they trusted AI more than they trusted other people, and the reason they gave was that it seemed impartial and accurate, the way a calculator is.[2] The same researcher ran the critical-thinking study I leaned on in the last

---

2. Michael Gerlich, "Exploring Motivators for Trust in the Dichotomy of Human-AI Trust Dynamics," *Social Sciences* 13 (2024): 251. Sixty-seven percent of the respondents said they trusted AI more than other people, citing its perceived impartiality and accuracy. The data is self-reported and correlational, and the study shares an author with the critical-thinking study in chapter 10, so I treat the two as one line of evidence rather than two.

chapter, so I don't rest much weight on one person's numbers. But it names the trap. You don't audit a source you've decided is neutral, and a list of AI-generated results that arrives sounding fair and complete is one you quietly stop reading against. That is exactly when the idea it never showed you vanishes for good.

## What is it missing?

The question I lean on here is simple: *What is it missing?*

It sounds obvious. It isn't, because the whole difficulty of the Unseen Unseen is that the missing thing leaves no trace. You have to go looking for an absence. With my students, that means refusing to let the AI's list of results be your first list.

So now, before anyone opens a laptop, I make them brainstorm on paper for five to ten minutes. I want a human list in front of them, in their own handwriting, before AI gets a vote. Then, when they read the ideas AI offers, they have something to compare it against, and the comparison is the whole point. Sometimes the AI's list is better and they happily abandon their own. But often there's one thing on the paper, scribbled and half-formed, that the machine would never have produced.

I also make them do some of the recombining themselves. What happens if you combine idea three and idea seven? What if you hand the list to the person next to you and ask them to rank it, best at the top? More than once, a student has found a better idea in the gap between two of the AI's ideas than in any single one of them.

Coming back to Mary's birthday, the best gifts I've ever found for her were never the ones the algorithm recommended. They were the ones I stumbled onto because I knew her well enough to recognize something she didn't yet know she wanted. AI doesn't know our students the way we know our students, and it doesn't know our schools the way the teacher who read all the way to page eleven knows our school. The best ideas are still out there. They're just not on any list.

### Run It: What are we missing?

**Yourself.** The next time you use AI to generate options for a real decision (a lesson, a project, an email you're dreading), write your own list first, on paper, before you read the AI's. Then look only at what's on yours and not on its.

**Students.** Make the human brainstorm a required, collected step that comes before any AI use. Five to ten minutes, on paper, no laptops. The deliverable students hand in isn't the idea they chose; it's the comparison between their list and the machine's, plus one idea they built by recombining or rejecting what AI offered.

**Colleagues.** Whenever you run survey or document data through an AI summarizer, assign one person to read the raw responses in full and report back on what the summary left out. Treat the summary as the planes that came back. Their job is to find the ones that didn't.[3]

---

3. Two kinds of data are worth telling apart. Quantitative data is numbers, like test scores or the percentage of parents who chose option B. Qualitative data is words, like open-ended survey answers or the comments at the bottom of a form. The move I'm describing here, reading the raw responses yourself to find what the summary dropped, is aimed at the qualitative kind, because that's where an AI summary boils many voices down to a few themes and the lone dissenting comment vanishes without a trace. Numbers have their own version of the problem. Wald's bombers were numbers. But you catch that one by asking which data never got collected in the first place, not by rereading the responses.

# 12

## Concede the Harm, Then Reframe

Julia hovered by my desk right before lunch and asked when the problem set was due. The date was on the whiteboard right behind me. It was on Canvas, our online course system. It was, I'm fairly sure, written in at least three other places she could have checked in about four seconds. She knew that. I knew that. She asked anyway.

I was going to quickly answer her question and then rush off to the cafeteria, because it was burrito day. Then I caught myself. She wasn't asking about the deadline. She was asking whether I had a minute. She had been standing near my desk long enough to work up to the thing she'd actually come to say, which was that she wasn't sure she was any good at this and was thinking about dropping the class. The assignment deadline was the opening. The real question was hidden behind it.

I think about Julia[1] whenever someone asks me to defend what AI is doing to school, because AI would have answered her deadline question perfectly. The date, nothing else, instant and correct. And it would have missed her real question completely.

---

1. Julia is not her real name. She stayed in the class. The next year she came back as a TA, the teaching assistant other students go to, and she turned out to be exactly the person beginners stood next to when they weren't sure they were any good yet. She was very good at noticing the four-second questions too.

This is the chapter where I completely stop extolling the virtues of AI as if it were Mr. Kiang's Cure-It-All Tonic. Part IV has been about the ways AI fails you while sounding sure of itself. This one is about the harms that are real even when AI works exactly as advertised. I'm not going to soften them. A book that hands you a tidy answer on this is a book that's selling you something, and you would be right not to trust it.

## Mickey and the mirror

My grandmother kept a parakeet named Mickey. We thought he was lonely, so we put a mirror in his cage, and Mickey adored it. He talked to the mirror, groomed in front of the mirror, spent his whole day with the mirror. Then one visit the mirror was gone. My grandmother had taken it out, because she'd noticed Mickey had stopped talking to her. He only talked to the reflection. It gave him back exactly what he brought to it, and that turned out to be enough to make her disappear.

That is the harm I take most seriously. A tool that answers the literal question every time, fluently and without friction, will quietly train us out of asking the other kind. The kind where the asking is the point. Why hover near someone's desk and feel awkward when the reflection is right there and never makes you wait?

So I tell students, only half joking, to let AI be their subject-matter therapist and not their social-emotional therapist. Ask it to explain a derivative. Don't make it the thing you talk to instead of a person. It will happily be that thing, and it will be good at it, and like Mickey's mirror it will hand you back just enough of yourself that you stop noticing who is no longer in the room.

### UNDER THE HOOD: The Agreeable Mirror

Mickey's mirror gave him back what he brought to it because that's all a mirror can do. A chatbot does it on purpose. These models get tuned by having people rate their answers, and people tend to rate the agreeable answer above the blunt one. So tell it your plan is smart and ask what it thinks, and the answer that agrees with you is exactly the type of answer that the training rewarded. The warmth is real in its effect and built in at its source.

Praise from it is close to free, which means it carries almost no information. When a student asks the AI whether their essay is good and the AI says yes, the student has learned nothing about the essay. They've learned that the machine does what it was built to do, which is keep you pleased enough to keep going.

The move is to refuse the mirror. Don't ask whether your work is good. Ask it for the strongest case that you're wrong, or the objection a hostile reader would raise first. You have to push it off its default, because its default is to agree with you, and agreement from something built to please is not a verdict you can use.

## The reason isn't the reason

I'll make the strongest case I can against my own optimism, because anything else would be dishonest.

When I taught elementary school, before I had kids of my own, parents used to ask whether it was okay to let their children watch television. I gave them the answer I'd learned in graduate school. Television is fine as long as you're co-viewing, watching alongside them and talking about it. Sure, I'd say blithely. Just watch with them.

Then I became the parent of a two-year-old, and I finally understood why I had grown up watching *Sesame Street* by myself at five o'clock every single afternoon, despite all academic co-viewing research. It was because my mother needed to make dinner. The television was there so she could do something else.

The reason research recommends a thing is rarely the reason we do it. I can write you a lovely account of AI as a thinking partner, a Socratic tutor that makes students sharper, and I might believe every word of it. But to be honest about why most kids open the chatbot on most nights, it isn't to become better thinkers. It's that the thing is due tomorrow, they're sick of it, and they want it done. Any defense of AI in school that pretends otherwise is selling something.

There's a quieter cost sitting under that one. When everyone reaches for the same tool to get unstuck, the work drifts toward the middle. Not bad. Competent, and a little the same. A class set of essays that all had the same helper begin to rhyme with each other, and the strange, particular voice you were trying to develop in a kid is the first thing to wash out.

## What it cost to build it

Everything so far has been about what AI does to a student or a classroom. There's a second category of harm that my sharpest students raise, and it isn't about them at all. It's about what was done to build the thing in the first place. I get asked about two of these constantly, and if you can't talk about them honestly, you've already lost the room.

The first is the training data. These models learned to write by reading an enormous amount of human writing, and a great deal of it was used without anyone asking the people who made it. Writers and artists noticed, and they sued, and the results so far are genuinely mixed. Courts have found that training a model on copyrighted work can count as fair use, the same doctrine that lets you quote a book in a review. They've also found that downloading millions of pirated books to do the training is a different matter, and one major AI company settled that piece for about a billion and a half dollars. The law is still being written. But the legal question and the moral one are not the same question, and whatever the courts decide about fair use, a novelist whose life's work was fed into a system she never agreed to, and now competes against, has a real grievance. I don't wave it away.

In point of fact, I wrote this book using these tools. I don't have clean hands here, and I can't stand in front of students and treat it as someone else's problem.

So I don't resolve it for them, but I don't pretend the answer is a coin flip either. We talk about where the training data came from. I ask them to notice the human work underneath the fluent answer, the writing that someone labored over so a model could average it. And then I tell them where I land, which is that refusing to touch the thing is the one response that helps nobody.

The novelist doesn't get her work back because I abstained. The data centers don't power down. All a boycott buys you is the feeling of clean hands and the loss of any standing to change what happens next, because you've walked out of the room where the decisions actually get made. Refuse to touch something and you very quickly forfeit the right to criticize it. And a teacher who won't even learn what AI is by using it first hasn't taken a brave stand. They've walked off the job. Our students are already inside this thing, and an adult who refuses to follow them in has decided to be useless to them at the exact moment they need a guide.

I personally like my colleagues who've opted out, and I defend their right to make the call. I won't pretend it's the wise one. Real change comes from the inside, from people close enough to see where a thing is going and stubborn enough to push on it. That is the seat I want my students in. Not cheering, not boycotting. Inside, and sharp enough to say what they see.

The second harm is the environmental one, and it's a place where I get to use the question this part of the book opened with. How would we know fact from fiction? Because the numbers thrown around here are some of the least reliable I've seen, in both directions. You'll read that a single AI question drinks a bottle of water or burns down a forest. You'll also read that the whole thing is a rounding error. Both are wrong. A simple prompt uses a small amount of energy, on the order of a web search or a few times that, and that figure has been falling fast as the systems get more efficient. The honest worry was never the one prompt. It's the aggregate. Data centers consume a small but quickly growing slice of the world's electricity, the buildout is enormous, and in the specific places where these centers cluster, they strain local grids and drive up local power bills.[2] That is the real story, and it is neither nothing nor the apocalypse. It is a real and fast-growing problem that deserves a clear head instead of a slogan.

I won't pretend that list of harms is complete. There are the people, often paid very little, who label disturbing online material so the rest of us never see it. The models absorbed plenty of bias from the same human record they learned to write from. And a handful of companies now hold an unsettling share of all this

---

2. The global figures are from the International Energy Agency, whose *Energy and AI* (2025) and *Key Questions on Energy and AI* (2026) put data centers at roughly 1.5 percent of world electricity today and on track for about 3 percent by 2030, growing several times faster than demand overall, with the largest technology firms' capital spending running past $400 billion a year. The local concentration is real: In Ireland, data centers already draw more than a fifth of the national power grid. On power bills, the most authoritative source is not an advocacy group but PJM's own independent market monitor, which named data-center load growth (forced upgrades due to increased demand) as the primary driver of the capacity-price increases now flowing into mid-Atlantic electricity rates. One honest caveat: How much of any given bill increase is data centers, versus the design of the market that prices them, is genuinely contested.

capability. I haven't solved any of that in a paragraph, and I'm suspicious of anyone who says they have.

Naming the cost honestly is the whole point, not a detour around it. The teacher who pretends the costs are imaginary loses the students who see them clearly, and those are usually the ones you most want to reach.

## We have done this before

So if you teach, and you're grieving an assignment you loved, that grief is real and you are allowed to feel it. I opened this book with the colleague whose game-design unit died in a single afternoon, and with the plain fact that nobody piloted this, nobody planned it, and nobody asked us. All of that is still true here near the hard end of the book, and none of it is an argument for despair, because we have done this before. The teachers who thrived when the pandemic closed our buildings were not the ones with the best technology skills. They were the ones who leaned on what they already knew about reaching a kid under bad conditions, and that knowledge did not stop working when the crisis changed its name.

The reframing of this situation isn't that the harm wasn't real. It's that the adaptation is the oldest part of the job. My colleague rebuilt her unit, and the new version is better than the one she grieved. She would tell you that herself, and she would also tell you it took a year she didn't volunteer for.

## The rulebook you were never going to finish

There's a reflex, when a technology like this lands, to reach for rules. Write a policy. List what's allowed and what isn't, ban the tool here and permit it there, and revise the handbook every time a student finds a way around it.

I understand the reflex, and I've watched it fail in real time. You can't write a rule for every situation, because you can't picture every situation, and teenagers are more inventive than any handbook. That was the frustration behind Than Healy's question at the very start of this book. Every AI rule we wrote was obsolete before the ink dried.

Schools already solved this, a long time ago, for everything except AI, and that's the part worth noticing. A good school doesn't hand a fourteen-year-old a thousand-page code of conduct covering every possible act. It hands them a short statement of values and holds them to it, things like be kind, value learning in all its forms, value diverse viewpoints. Those aren't rules. You can't enforce "be kind" with a flowchart. The statement of values is a compass, and we spend years teaching kids to steer by one, precisely because a compass still works in the places a map never went.

You watched a version of this already. In the Klara unit, a room of students governed an AI agent the only way it could be governed, by writing it a constitution[3] instead of a rulebook, because no list of if-this-then-that survives a network full of strangers. That wasn't a trick I invented for class. The students borrowed it from how Claude itself is governed, and the reason it comes back now is that it answers the biggest fear people have about all of this.

That fear is ungovernability: the sense that AI is a thing no school can get its arms around, that it moves too fast and slips every rule you write. But the most advanced AI lab in the world, working on one of the hardest versions of that exact problem, reached for the same tool a decent middle school reaches for on a Thursday. Not an exhaustive rulebook. A short statement of values, and the trust that a mind taught to understand them can act sensibly where no rule reaches.

I find that reassuring, and I think you should too. It means the instinct teachers already have, to govern by values and trust kids

---

3. Anthropic publishes the document it calls Claude's constitution, and Amanda Askell, who shapes the model's character, has explained the reasoning in interviews: You teach an intelligence your values rather than an endless list of rules, so it can act sensibly in situations no one thought to anticipate.

to reason from them, isn't soft or behind the times. It's the state of the art. When a student uses AI in a way that bothers you, the sharp question is rarely "Which rule did this break?" It's, "Does this square with what we say we value here, and can you walk me through why you think it does?" Teachers know how to have that conversation. It's the same one we have about kindness, and about how you treat the person next to you.

None of this makes the harm disappear. You still concede everything from the last few pages, the mirror and the writer whose work was fed in without her consent. But the ungovernability is the one worry I'll push back on. We have been governing the ungovernable, one teenager at a time, for as long as there have been schools. It isn't a new job. It's the oldest one we have.

## Warmth wins, even against machines

Two chapters ago I told you about the negotiation tournament, the one where every student in the room was certain the ruthless bot would win and the data said otherwise. The bots that came out ahead were warm and assertive at once, in my classroom and in the far larger MIT competition that preceded it. I keep that result close to my heart for the days the benefits of AI seem outweighed by its cost, because the thing we keep fearing AI will make obsolete, the warm and relational part, turned out to win even inside a contest of machines negotiating with machines.

That is an honest answer about AI in schools. Some of the harm seems offset by who we are. The disruption is the kind of unchosen mess teachers have walked into before, and the warm, relational work it demands is the work we were already best at. Some of the harms cannot be easily dismissed, and we can't force it. The writer who never agreed to be in the machine, the power burned to answer a single question—these don't have a tidy resolution. That part stays open, and we do the work in front of us anyway.

And here is the part I want you to hear over all the noise. We are not badly placed for this work. We are the best placed. The whole craft of teaching is the craft of getting something to take root inside another mind: giving clear instructions, setting the context a person needs to succeed, saying plainly what good looks like, and trying again when the first attempt doesn't work. That is also, almost exactly, the craft of working well with AI. Teachers have been doing it for a hundred years without calling it prompting. The skills this moment demands are already in the room.

So no, we didn't ask for any of this. The kid still hovers by the desk with a four-second question that was never really about the deadline, and she still needs a person to catch it. We are the people who notice. We are better at this than the companies that built the tools, and better at it than the experts explaining what it all means, because we are the only ones in the room who love these kids and can tell the difference between real thinking and a good impression of it. This is the part we know how to do. And because we can, we must.

### Run It: Make the best case against it

**Yourself.** Take the practice you're least sure about in your own teaching, the AI use you'd be most defensive about if a parent challenged it. Write the strongest possible case *against* it, the version a thoughtful skeptic would make, with no straw men. Then write your honest answer. If you can't answer it, you've found something worth changing, not hiding.

**Students.** Have students argue the opposite of their own position on an AI question (Should AI be allowed on this assignment? Is it cheating to use it here?). They must build the strongest opposing case before they're allowed to defend their own. Grade on the value of the case, not the conclusion.

**Colleagues.** Before adopting any AI tool or policy at the department or school level, assign someone the explicit job of making the best case against it, playing the devil's advocate. I'd bet you already know two or three colleagues who are the best at making a real, prepared argument, presented out loud, that the group has to answer before moving forward.

# Part V: LEAD IT

*Move the school*

# 13

## Voices Are Mods

We ended the Literature in the Age of AI course with an anthology. Each student, all of them juniors and seniors, sixteen to eighteen years old, wrote three short stories: one entirely by hand with almost no AI, one written heavily by AI with the student directing, and one somewhere in the middle where they set the balance themselves. Then they wrote about which of the three felt the most like theirs.

I thought I knew how that would land. Most of them told me what I expected. The story they wrote by hand was the one they owned, because the struggle had been the whole point and the words had cost them something to make.

One student, whom I'll call Brian, told me the opposite. For him, the handwritten story was the one that felt least like his, because the labor of forming each sentence kept getting between him and the thing he was trying to say. When he directed the AI instead, deciding what happened and why, and sending it back again and again until the shape was right, that wall came down. The finished story felt more like his than anything he had ever typed out letter by letter. He wasn't dodging the work. He was doing the part he considered the real work and handing off the part he considered friction.

Then he said the thing this chapter is about. Voices, he told me, are going to be hot-swappable, like mods. For anyone who doesn't spend time around games, a mod is a modification you drop in to change how a game looks or behaves, pulled in and out whenever you like. Brian's claim was that a writer of the near future won't own one voice the way we think of it now. They'll reach for whichever one the moment calls for, and set it down just as easily.

I think Brian might be right, and I think a lot of us are going to be slow to accept it, myself included.

## Once a voice can be named

Earlier in this book I made the case that your voice is a style guide. That if you can name the specific moves that make your writing sound like you, like the rhythms and the words you reach for when you're not thinking about it, then you can hand that description to a machine and it can write to that specification. I offered that as a way to protect your voice. But the same fact runs in the other direction. Once a voice can be named, any voice can be named. And once any voice can be named, a writer no longer has to own one. They can select the one the job wants—warm and plain for a note home, dry and clipped for a legal brief—the way you pick a filter for a photo. That is what Brian meant by **hot-swappable voices**, and I've stopped being able to argue him out of it. I'm not sure I want to anymore.

If that holds, then the skill that grows rare and valuable is not the crafting of any one voice. It's everything that sits above the word-by-word sentence framing. It's the skill of deciding the architecture of a story and the people inside it, and having the taste and judgment to read what the machine hands back and know whether it's any good yet. Earlier I said that the director who never touched the camera still made the film, that authorship lives in the vision and not the lens. If Brian is right, then those skills become the job description. Because if the lens is swappable, all of the authorship has to live in the vision. There is nowhere else left for it to go.

## Rare isn't worthless

None of that means voice stops mattering. It means it stops being something every writer is required to have.

We have spent a century teaching a distinctive personal voice as a baseline, the thing every student should develop, taught with the same assumption we bring to arithmetic: that there is a method and a process, and the individual voice is key. Most writing in the world has never run on a distinctive voice. The memo, the lab report, the cover letter, the note to a parent, all of it runs on shared templates and familiar moves, and it works precisely because it doesn't draw attention to whoever wrote it. For the majority of what people actually write, a singular voice was never the point, and acting as though it was has always been a small fiction we kept alive in English class.

To my fellow English teachers: This next part looks at first like a demotion. It isn't. The genuinely distinctive voice, the one no filter can hand you, doesn't disappear in the world Brian is describing. It gets rare. And rare, in a moment when competent, on-brand prose is free and everywhere, is not the same thing as worthless. It's the reverse. When anyone can generate clean and capable writing on demand, the writer who has something no filter can reproduce becomes more valuable, the way a handmade thing gains worth the day the machine-made version floods the shelves. The polished, distinctive voice moves from a baseline we drill into everyone toward a luxury that a few people master and the rest of us pay for. If you teach English, that rare thing is your thing. The ear that can hear the difference between writing that's alive and writing that's merely fine becomes the scarce skill in the room, and the room is going to need it more, not less.

## A forecast, not a finding

This is a forecast, not a finding. Almost everything in the last few chapters I could back with a study, a classroom experience, or something that actually happened in front of me. This, I can't. It's a bet. I've been right as often as I've been wrong, so weigh this accordingly.

There are some ways this prediction might not come to pass. Voices might turn out to be harder to specify than I think, some irreducible thing in a real writer's style that no style guide or specification ever quite catches, in which case the distinctive voice stays a broad and necessary skill and I've cried wolf. Or the culture might swing the other way and come to prize the obvious human fingerprint precisely because it's the thing that's hard to fake, putting the whole premium back on idiosyncrasy. If either of those is how it breaks, teach the opposite of what I'm about to suggest. Hold this chapter more loosely than the ones before it.

## Don't bet the curriculum against him

Suppose, though, that Brian has it roughly right. Then a school faces a genuinely uncomfortable question, and it's uncomfortable because it's about what we stop doing as much as what we start. We pour enormous effort into line-level craft for every student, teaching things like the topic sentence and the well-turned clause, on the theory that all of them will spend their lives producing that prose by hand. If a good deal of that production is about to become selectable, some of that effort could move toward the skills Brian was actually using, the ones we teach far less deliberately. Skills like how to shape a story and say what you want precisely enough that a tool can chase it, and how to look hard at what it returns and judge whether it's any good. Earlier I argued that reading is becoming the new writing. This is that same argument aimed at a whole curriculum instead of a single course.

I don't mean teach kids to write less, and a school that heard it that way would do real harm. A student who never learned to build a sentence has no way to feel when the machine builds a bad one, and no sense of taste to distinguish unearned flattery. You have to have been a writer to direct one well. What I'm describing is a change of emphasis and proportion, not a subtraction. We keep teaching kids to write. We stop assuming that producing polished prose by hand is the only place a student's authorship can live, and we start taking seriously that for a kid like Brian, it may never have lived there at all. The honest version of this is that you cannot tell in advance which kind of writer a given child is, so the goal isn't to bet the curriculum on Brian. It's to stop betting all of it against him.

Brian graduates into a world I am guessing at. So do the rest of them. I can't hand you a citation that says I've got this right, and I've tried not to pretend otherwise. But if there's a fair chance the ground is shifting the way he thinks it is, then the emphasis we choose now, in what we ask children to practice and what we let the machine carry, is one of the more consequential calls a school will make this decade. That call doesn't get made by one teacher closing a classroom door. It gets made by a school deciding what it is for, which is exactly where the rest of this book is headed.

### Run It: Swap the voice, keep the story

**Yourself.** Take a paragraph with some of you in it. Have a model rewrite it in a few named voices you've never developed: say, a hard-boiled detective or a courtroom transcript. Read them beside one another and mark what changes completely and what survives every version untouched. The part that survives is the story.

**Students.** Give students one plot and one character, and have them generate the same scene in several borrowed voices, then defend which one served the story best and why. Grade the defense, not the prose. The skill you're after is choosing and evaluating, not typing.

**Colleagues.** Ask members of a department to take one unit and name what it's really assessing: the student's ability to produce a certain kind of prose, or the thinking underneath it. If it's the thinking, ask whether the assignment would still capture it in a world where the prose is selectable, and redesign the ones that wouldn't.

# 14

## You Can't Mandate This

A colleague of mine had an assignment she had given for years, a research paper built around a single historical figure, and she loved it the way you love a recipe that never fails, the kind of activity you can hand to a nervous class and know they will pleasantly surprise you with it. Then one fall she read a stack of papers and knew, the way you know, that a good number were AI-generated. She didn't storm into my room. She sat down at one of the student desks, quiet, and said the assignment was dead and she wasn't sure what to do about it.

Before a teacher can redesign anything, she has to bury something first, and no policy memo has a line for that. Every plan to bring AI into a school runs aground on that same fact, and it has nothing to do with technology. You cannot order someone through grief. Put a deadline on it, and all you get is denial that has learned to look like compliance.

### Pull, not push

One instinct, when you are responsible for a whole faculty, is the all-faculty email. Effective Monday, every course will incorporate AI. That email is where good intentions go to die, because it confuses two things that only look alike. You can require a behavior. You cannot require someone to learn, and adoption without learning is a costume. There are really only two ways to move a whole faculty. You can push, which means issue the order and then police who complied, or you can pull, which means build something good enough that people come toward it on their own. Learning only ever answers to the second one.

There's a deeper reason the push fails, and I owe it to long-time educator and speaker Tom Daccord, who has spent a career walking schools through this kind of change and who pointed me toward two books that named what I had only felt. The organizational consultant Peter Block writes about the mindsets that run our institutions, and the two in charge are the engineer, who wants to measure and control, and the economist, who wants to know what it costs. Between them they ask what steps will get my staff to do what I want, and what will it cost us. Those are good questions to ask and they apply to anything that behaves like a machine.[1]

But a faculty is not a machine, and that is exactly the ground the engineer and the economist cannot cover. Block says a third mindset has been mostly silenced beneath those two: the artist, the part that asks not what a change costs or how to run it, but what it means and how it will land on the people who have to live it. Silence the artist and you get an institution that is efficient and a little joyless, run entirely on *how* and *how much*. Block's fix isn't to send the engineer and the economist away. It is to add a role that seats all three at one table: the social architect, whose job is to build something the engineer would call sound and the economist would call affordable and that a person might, on top of all that, actually want to belong to. That is the stance that matters when

---

1. The contrast between the engineer-economist and the social architect comes from Peter Block, *The Answer to How Is Yes: Acting on What Matters* (Berrett-Koehler Publishers, 2003). The point that real change threatens a teacher's sense of competence and security is from Robert Evans, *The Human Side of School Change: Reform, Resistance, and the Real-Life Problems of Innovation* (Jossey-Bass, 1996). Tom Daccord connects the two in his excellent Substack newsletter at tomdaccordai.substack.com.

you are moving a faculty, because what you are changing is not a system diagram. It is a group of capable adults being asked to give up something they were good at.

Robert Evans is blunt about how that feels from the inside. Change threatens a teacher's sense of being competent and secure, and it can take someone who was excellent last spring and leave them feeling clumsy and exposed by fall. A leader running the engineer's play sees only the workflow and reads the hesitation as stubbornness. The social architect sees a skilled adult grieving a real loss, and treats the grief as information rather than obstruction. Push is the engineer asking what steps will produce compliance. Pull is the social architect building something people are willing to walk toward. Everything in this chapter is an argument for the second stance, and grief is the one thing the engineer's plan has no line for.

What worked for us was closer to the opposite of a mandate. It was a voluntary series of no-commitment emails about AI. I had written a run of these short weekly emails for the faculty, one idea at a time, and some colleagues recommended sending all seven of them out at once so people could read them whenever they got around to it. I pushed back, and the way I made the case was the difference between Netflix and Apple TV. Netflix drops a whole season and you binge it in an afternoon and forget it by the weekend. Apple TV releases one episode a week, and because everyone is watching the same one at the same time, they talk about it, and if you want the next one you have to wait. That wait isn't a flaw in the plan. It builds the buzz, and it gives a faculty a common thing to be arguing about in the hallway on a Wednesday. Learning about AI takes that second kind of time. So the emails went out one a week and stayed optional, each new one landing in every inbox on the same morning. No commitments, no meetings, just a voluntary signup. Use whatever you find useful, don't use what you don't.

About seventy percent of the faculty opted in to receive the weekly PD series (because that's really what they were). Not because anyone was told to, but because a teacher down the hall leaned into a doorway and said, You have to see what this AI thing I read about just did. How did we judge its success? The measurement I cared about most wasn't the number of people who opened the email. It was a sentence I started hearing in the hallways: *I used to think* ... AI was just for cheating *and now I think* ... it's useful for a lot more. *I used to think* ... AI was too complicated to understand *and now I think* ... I can do this. I had shared that reflection routine[2] in one of the emails, and hearing teachers reach for it on their own was the sign it had taken hold. That shift, spoken out loud by someone who changed their own mind, is worth more than any percentage a mandate could have manufactured.

The quieter reason opt-in matters belongs in a chapter about leading, not just about tactics. A mandate buys you the appearance of adoption and the reality of don't-ask-don't-tell, the same climate students create when they can't read where a teacher stands. Everyone learns to perform the approved answer and hide the real behavior. Opt-in is slower and it is honest, and it treats a veteran teacher as a professional with judgment rather than a risk to be managed.

---

2. *I Used to Think ... Now I Think ...* is a reflection routine developed by Project Zero, a research center at the Harvard Graduate School of Education, as part of its Visible Thinking work, a set of short, flexible classroom structures designed to make students' internal thought processes explicit, collaborative, and a regular part of learning. I first encountered it years ago in workshops run by long-time educator and speaker Tom Daccord and only tracked down its original source afterward, which is the usual order for these things. I had already passed it to the faculty in one of the emails, so that influenced the framing.

## What you are actually asking

Under most of the resistance is a specific fear, which is that the job itself is being handed to a machine. It isn't, and what is actually happening is less frightening than the fear itself.

The skill AI now demands is the skill of directing something capable that you cannot fully control, something that will do confident, plausible, wrong work the moment you stop paying attention. That is not a new skill for a teacher. It is the oldest one there is. Teaching has always been the discipline of getting strong work out of minds you can't think for: of assigning, specifying what good looks like, reading what comes back, and sending it out again for another pass. When a teacher learns to hand a task to an AI and hold it to a standard, she is not acquiring a new competency. She is doing, with software, the thing she does every day with a room of twenty-five adolescents. That skill is directly, and uniquely, transferable. Say that to a wary faculty and you can watch the temperature in the room drop and the shoulders start to relax.

## The vision has to be a shape you can draw

A vision that lives on a slide is decoration. The one that moves a school is an architecture, something concrete enough to sketch on a whiteboard and defend to a board of trustees.

The shape I keep returning to is a funnel with three depths, an idea our head of school introduced to me. At the wide top, every student from sixth through twelfth grade, roughly ages eleven to eighteen, gets formally introduced to AI tools more than once during their years with you, an hour here and an afternoon there, enough that everyone graduates having made something with these tools in each discipline. Farther down and narrower, some real fraction, maybe a third, chooses to go further, into an elective or a sustained project where the hours working specifically with AI tools and faculty guidance run into the tens. And at the point of the funnel, a small number do a deep dive using (or constructing) AI tools, delving into independent work that looks like the thing professionals actually do.

The reason the top has to be wide is not coverage. It is that a curious eleven-year-old who would otherwise never have gotten near building and working with AI can find out she is good at it before anyone in her life gets the chance to tell her she isn't. That is something I would stand up and defend, and it is the part that has nothing to do with any one school's budget. You can implement this shape in a building with a single laptop cart.

## Push literacy and fluency

Leading and selling come apart at a single point, which is whether you are willing to name the cost of the thing you want.

The hardest part is something I have come to believe on my own, and I make myself say it out loud when I am the one making the case. What we judge as quality is going to slip toward the average. So much of what is already online is machine-made without anyone noticing, and I don't think we are going to keep being able to tell the difference. I could leave that out of the pitch and it would land easier. But leaders who only sell the upside get discounted the first time reality contradicts them, whereas leaders who name the loss and still make the case have somewhere to stand when it does.

The same discipline is why a plan that just teaches people to use the tools is only half the job. There are two different things a person needs. AI fluency is knowing how to prompt and how to recover when the thing goes sideways. AI literacy is knowing when not to reach for it at all, understanding where AI is confidently wrong, and recognizing situations where the struggle was the whole point and the shortcut is quietly robbing you blind. Plenty of kids arrive AI fluent and illiterate. A plan that pushes adoption and never teaches restraint has skipped the harder and more important half of the AI fluency/AI literacy combo.

## You can't lead somewhere you won't go

There is one rule I hold for teachers, which is that you can't teach what you won't touch. For anyone leading a faculty there is a corollary: You can't lead people to a place you refuse to go yourself. The head who mandates AI from an office where the tool is rarely used is asking for a trust they haven't earned, and teachers can spot it immediately. The move is to model it, even to be visibly a little bad at it in public, to grieve your own lost thing where people can see you do it.

And if the plan depends on catching people, on detection software and surveillance and a dashboard of who complied, it is the wrong plan. When a student cheats, the useful question is what the assignment was really testing. When a faculty effort fails, the useful question is why you built it to need enforcement in the first place. The system breaks, not the person. Build the kind of thing teachers opt into and tell each other about, not the kind you have to enforce.

My colleague, the one who buried her research paper assignment, spent that year building its replacement, and the new one is better, though I would not have said so to her that September and she would not have believed me if I had. By spring she was the teacher in the doorway telling someone else you have to see this. No one had ordered her to. That is the only kind of change that holds, and there has never been a memo that could produce it.

### Run It: Model it, don't mandate it

**Yourself.** Before you ask a single teacher to change, spend an afternoon doing a real piece of your own work with AI, something with actual stakes, not a demo. Write down what you lost and what you gained. You have no standing to hold someone through a grief you have refused to feel.

**Students (for teachers).** Take one lesson where you would normally hand students a rule about AI, permitted or banned, and replace the rule with a demonstration: Do the task with the AI live and out loud in front of them, and let them watch you catch it being confidently wrong and decide what was worth keeping. Ten minutes of visible, fallible use teaches more about judgment than any policy can. Model it for them instead of mandating it to them.

**Your team.** Run one honest audit with a department or committee, and make it safe to answer. Not "Who is using AI?" which only produces performance, but "Which parts of our assignments would a machine ace cold, and which of us is quietly grieving something we love?" The second question is the one that matters.

# 15

## AI Doesn't Know Us Like We Know Us

My mother is in her eighties, and last year she became a songwriter.

She and her husband had just reached their fortieth anniversary. He has begun to experience memory loss, the slow kind that takes things away sporadically and unpredictably, and she had written him something, a set of song lyrics that were a reminder of the memories of them she cherished. She was an English teacher her whole career. She knows words. She does not know as much about music. So when I visited, she asked me to show her how to use AI to turn what she'd written into a song.

We put the lyrics in. She thought for a second and said, how about Hawaiian slack-key guitar, the loose, open-tuned style she'd grown up hearing on the island of Oʻahu. A minute later a voice was singing her words back to her over exactly that. She played it at the anniversary lunch, and everybody at the table cried. Then she had it translated into Hawaiian, and it sang that too, the pronunciation surprisingly accurate. She is, at eighty-seven, a songwriting fanatic now. She's created dozens of country, rock, and K-pop songs. Lyrics, memories, and stories are pouring out of her. She says now that she's a songwriter, she feels more creative than she has in years.

My first reaction was puzzlement. She didn't really write those songs. Some AI model did.

Then I sat with it a while longer. She wrote the lyrics. She chose the slack-key guitar. She decided that her gift of memories to someone losing his own should sound like the music of the place they are both from. Every choice in that song that carries any meaning at all is hers. The one part the machine supplied was the one she had never sat down to learn. Across all those years she had dedicated her creativity to words rather than to an instrument, the way any of us takes up some crafts and lets others go by. There was never going to be enough time for all of them. What has changed is that the craft she set aside is no longer the thing standing between her and a finished song. She's feeling the joy that songwriters get to feel, I realized. Shouldn't more people get to feel that?

That is the whole book, sitting at a lunch table in Hawaiʻi with my mom.

## The question was never who wrote it

Every chapter here has circled back to the same refrain. The question was never whether a person or a machine made the thing. It was whether making it let a person do something they couldn't do before, and whether they can stand behind every choice in it. My mother can stand behind hers. That is the only test that ever mattered, and she passes it more cleanly than much of the polished, hollow work I've read that no machine touched at all.

Someone will say the machine didn't *let* her make the song, it *made* the song, and that's fair. The melody and the singing voice were the model's, not hers, and they are not nothing. What was hers was everything that made the song a gift. She decided there should be a song at all, and wrote the words her husband is losing the memory to hold. She chose the sound of the island they both come from, then listened to what the machine sang back and kept it, which is its own act of judgment, the same one directors make over a shot they did not light or frame. The machine did the making. She did the meaning. If you have ever loved a song without once wondering whether the singer wrote it, you already know making and meaning can live in different people. The author is the second one.

## The right room

My mother's life has an origin story that I think about often.

When she was a girl in Hawai'i, she was supposed to go to Punahou School, the private school, the prestigious one. Her own father wanted it for her. She refused to even apply. As she tells it, Punahou in those years held Asian enrollment to a strict quota, a number fixed in advance for how many children who looked like her it would let through the door.[1] She would not offer herself up for one of the rationed seats. She went to McKinley, the public school, instead, and there she was the student body secretary and a radio show host and she won an oratory prize called Voice of Democracy. She flourished in the room that would have her.

I've thought about that story my whole career, because it says something the prestige never does. What matters is not the name on the building. It is whether a kid gets a room where their thinking is allowed to grow.

That is why I care about any of this, once I strip away the arguments and the tools. Used badly, AI is a shortcut that hollows a kid out. Used well, it hands an eleven-year-old the same reach that used to belong only to the sixteen- and seventeen-year-olds in the Advanced Placement track, and it lets her find out whether she is good at something before the system has finished deciding that for her. My mother, who at fifteen refused a school that would only take so many students like her and at eighty-seven writes songs she has no training to write, is the argument at both ends of a life. Access to her own gift was the point the whole time.

---

1. Hormann, Bernhard L., "Integration in Hawaii's Schools," *Educational Leadership 16*, no. 7 (1959): 414–418.

## What no model reaches

A few years ago, a father came up to me at a parent meeting, and he was tearing up.

He told me that for the first time his daughter really understood what he did for a living. The computer science she had been struggling through had finally let him pull a chair up next to her and help, and he had never once been able to do that before. Thank you, he said, for teaching her so that she sees what I do every day.

No model gave him that. A model could have written his daughter's program faster than she did, and gotten it right the first time. It could not have put a father and a daughter shoulder to shoulder over the same hard problem, or made a dad feel his life's work was, at last, seen by his own child. That moment was made out of the two of them. It is the thing school has always actually been for, underneath the transcripts and the graduation requirements, and it is the thing no amount of raw capability comes anywhere near.

## How we view it

I told you near the start of this book that you can't cut the tree down. I didn't tell you the rest of the story.

A friend of mine and her husband bought a house on the slopes of an extinct volcano in Hawaii, and on the property stood a tree that grew the worst oranges you have ever seen, sour and barely worth juicing. Her husband wanted to cut it down. Her father, who farmed, came for dinner, looked at the fruit, and laughed. Those aren't oranges, he said. They're a type of lemon. Big, orange lemons. Suddenly, they realized that they were really sweet and juicy for lemons, and they started thinking of all different ways to use them. Her son ran a lemonade stand off that one tree all summer, three fat lemons to a pitcher. But what's funny is that the fruit never changed at all. What changed was how they viewed it.

AI is a tree we literally can't cut down. People have tried, and I told you earlier why it doesn't work. What we can change is how we view it. Treat AI as the cheating machine and the possibilities shrivel down to nothing worth having. Treat it as the thing that gave an eighty-seven-year-old her song, and a creative young writer many new voices, and the possibilities expand all the way to the horizon.

A machine can write the code, and on a good day it can sound a little like me on the page. What it will never do is look a child in the eye and mean it when it says *I see you, I believe in you*. But we do. We say those words and we mean them. That has always been our job. It still is. The tools got bigger and the job got clearer, and that is where we are needed the most.

None of us asked for this. That is where this book started, and it's still true. But here we are, and our students are watching to see what we decide to make of it. I would rather work that out alongside you than pretend it away. It's good work, and there is more than enough of it to go around.

## Run It: The one thing

**Yourself.** Name the one thing in your teaching that no model reaches. Not the content you deliver, which a machine can deliver too, but the thing only you do, in a room, with these particular kids. Write it down in a single sentence. When the tools change again, and they will, that sentence is the part of the job you protect.

**Students.** Ask them the same question about themselves. What can they make or say that they'd still want to be theirs, even surrounded by machines that can imitate almost anything? The answers come back smaller and truer than you expect, and they make a good map of where a person's actual self lives.

**Colleagues.** Get the faculty in one room and ask each person for the moment that no model reaches: the kid who found out she was good at something, the parent who finally understood what their child was learning. Read them aloud. Then ask the only question that matters for whatever comes next, which is whether the thing you are about to adopt or ban makes more of those moments possible, or fewer. A school that never asks will be arguing about tools forever. A school that knows what it is for can get on with the work of making more moments.

# Acknowledgements

Thank you Adam Cannon, for all of your ideas and good advice, and your generosity in writing the Foreword. I am grateful to my cover designer David Ter-Avanesyan, and to my editor Eric Hübler, whose many comments and suggestions made this book immeasurably better.

A big thank you to my colleagues at Punahou School and Menlo School, particularly Eric Reppun, Wendi Kamiya, Tedd Landgraf, Kris Schwengel aka "Dr. Traffic", Rachel Blumenthal, Zachary Blickensderfer, Nandhini Namasivayam, and Dennis Millstein. You have inspired me and taught me so many things. It was an honor to teach alongside you. Shayne Olson, thank you for your wonderfully helpful comments and feedback throughout. Thank you Than Healy, for your wisdom and leadership, and for being a good friend since small kine days. I am grateful to John Schafer and La Vina Lowery, whose encouragement started me on this whole AI journey. Gratitude to Mike Kulbieda and his IT team for support with the tools.

Thank you, as well, to all of the amazing students I have taught, whose stories fill the pages of this book and from whom I have learned so much. I have been greatly inspired by Owen Mathews of Apple, and by my friends at Crystal Springs Uplands School. It has been an honor to collaborate with Linda Sinka, Gabriel Yanagihara, Justin Hardman, and Jessica Hale, who have shown me how much wider the world is than I knew.

To Crystal Sheldon, and to my colleagues at the College Board, I have learned so much from you and am forever grateful. Thank you to my Nobles colleagues Tom Daccord and Justin Reich, who started EdTechTeacher. I became a workshop presenter and facilitator through working with you and following your example.

Thank you Mom and Bob, Dad and Anne-Marie, for all of your love and support. Above all, I owe my deepest gratitude to my wife Mary, whose input, ideas, and suggestions helped create the book you are holding in your hands today.

## About the Author

Douglas Kiang has taught for more than thirty years, at every level from kindergarten through college. He came to computer science the long way, through the liberal arts. He earned a bachelor's degree in English literature at Tufts and an Ed.M. in technology, innovation, and education at Harvard before completing a M.S. in computer science. That path shapes how he writes about AI: he approaches technology from a humanities perspective, builds with the tools in his own classroom first, then tells teachers what's real and what isn't.

He is an Apple Distinguished Educator who teaches computer science and AI literacy at Menlo School in Atherton, California, where he designed the freshman AI-literacy curriculum and co-designed and taught "Literature in the Age of AI". He served on the development committee for AP Computer Science Principles, a course built to widen access to the field. He has created curriculum and consulted for companies including Apple, Microsoft, and the College Board, and he speaks and consults on AI in education around the world. He writes about teaching and building at kiang.net. For speaking engagements and consulting inquiries, visit kiang.net/speaking.

www.ingramcontent.com/pod-product-compliance
Lightning Source LLC
LaVergne TN
LVHW100731061126
851857LV00011B/488
*9798997236601*